INTO THE
DEEP

INTO THE DEEP

A MEMOIR FROM THE MAN WHO FOUND *TITANIC*

ROBERT D. BALLARD

and CHRISTOPHER DREW

NATIONAL GEOGRAPHIC

Washington, D.C.

Published by National Geographic Partners, LLC
1145 17th Street NW Washington, DC 20036

Financially supported by the National Geographic Society.

Library of Congress Cataloging-in-Publication Data
Names: Ballard, Robert D., author. | Drew, Christopher, 1956- author.
Title: Into the deep : A memoir from the man who found Titanic / Robert D. Ballard and Christopher Drew.
Description: Washington, D.C. : National Geographic, [2021] | Includes bibliographical references and index. | Summary: "Oceanographer and marine biologist Robert D. Ballard looks back on a long and storied life that includes accomplishments ranging from discovering new life-forms to finding the wreck of the Titanic"-- Provided by publisher.
Identifiers: LCCN 2020044366 (print) | LCCN 2020044367 (ebook) | ISBN 9781426220999 (hardcover) | ISBN 9781426221002 (ebook)
Subjects: LCSH: Ballard, Robert D. | Oceanographers--United States--Biography. | Marine biologists--United States--Biography.
Classification: LCC GC30.B35 A3 2021 (print) | LCC GC30.B35 (ebook) | DDC 551.46092 [B]--dc23
LC record available at https://lccn.loc.gov/2020044366
LC ebook record available at https://lccn.loc.gov/2020044367

Since 1888, the National Geographic Society has funded more than 14,000 research, conservation, education, and storytelling projects around the world. National Geographic Partners distributes a portion of the funds it receives from your purchase to National Geographic Society to support programs including the conservation of animals and their habitats.

Get closer to National Geographic Explorers and photographers, and connect with our global community. Join us today at nationalgeographic.com/join

Interior design: Nicole Miller

Printed in the United States of America

21/VP-PCML/1

To Barbara

CONTENTS

Prologue 11

Chapter 1 | FINDING NEMO 13

Chapter 2 | SWISS ARMY KNIFE 27

Chapter 3 | BECOMING A SCIENTIST 41

Chapter 4 | REWRITING THE SCIENCE BOOKS 61

Chapter 5 | MAKING FRIENDS WITH THE NAVY 79

Chapter 6 | MAY GOD BLESS THESE FOUND SOULS 99

Chapter 7 | TAKING STOCK AFTER *TITANIC* 119

Chapter 8 | MY CINDERELLA STORY 139

Chapter 9 | LAID LOW 157

Chapter 10 | A NEW PARTNERSHIP 173

Chapter 11 | ROLLING THE DICE 189

Chapter 12 | BLACK SEA QUEST 211

Chapter 13 | CLOSE-UPS ON THE PAST 227

Chapter 14 | BECOMING CAPTAIN NEMO 247

Chapter 15 | ALL THAT *NAUTILUS* COULD DO 263

Chapter 16 | DISCOVERING MYSELF 281

Chapter 17 | SEARCHING FOR AMELIA 293

Epilogue 307

Expeditions, Publications & Media 311

Acknowledgments 315

Source Notes 319

Index 325

Illustrations Credits 334

ROBERT D. BALLARD EXPEDITIONS

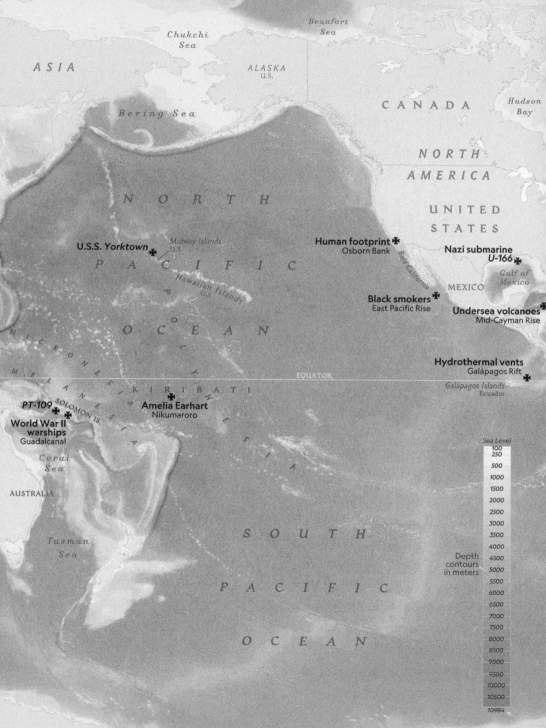

ARCTIC OCEAN

Chukchi
Sea

Beaufort
Sea

ASIA

ALASKA
U.S.

CANADA

Hudson
Bay

Bering Sea

NORTH
AMERICA

UNITED
STATES

N O R T H

Midway Islands
U.S.

U.S.S. Yorktown ✠

Human footprint ✠
Osborn Bank

Nazi submarine
U-166 ✠

P A C I F I C

Hawaiian Islands
U.S.

Gulf of
Mexico

MEXICO

Black smokers ✠
East Pacific Rise

Undersea volcanoes ✠
Mid-Cayman Rise

O C E A N

EQUATOR

Hydrothermal vents
Galápagos Rift ✠

Galápagos Islands
Ecuador

K I R I B A T I

PT-109 ✠
SOLOMON IS. ✠

Amelia Earhart
Nikumaroro ✠

World War II
warships
Guadalcanal

Coral
Sea

Sea Level
100
250
500
1000
1500
2000
2500
3000
3500
4000
4500
5000
5500
6000
6500
7000
7500
8000
8500
9000
9500
10000
10500

Depth
contours
in meters

AUSTRALIA

S O U T H

P A C I F I C

Tasman
Sea

O C E A N

ANTARCTICA

2000 mi
2000 km

10984

ARCTIC OCEAN

GREENLAND

Greenland Sea

Barents Sea

Baffin Bay

Norwegian Sea

Labrador Sea

Island of Newfoundland

Loch Ness ✠

North Sea

UNITED KINGDOM

Baltic Sea

(ÉIRE) IRELAND

EUROPE

War of 1812 warships
Lake Ontario ✠

R.M.S. *Lusitania* ✠

Celtic Sea

Black Sea ancient ships Sinop

Ph.D. research on seabed
Gulf of Maine ✠

Bismarck ✠

World War I wrecks
Gallipoli

Black Sea
✠

R.M.S. *Titanic* ✠

Isis
Skerki Bank ✠

GREECE TURKEY

Turkish jet pilots

Mid-Atlantic Ridge ✠

CYPRUS ✠ SYRIA

TUNISIA

Mediterranean Sea

ASIA

NORTH

Eratosthenes Seamount

Caribbean Sea

ATLANTIC

Red Sea

OCEAN

AFRICA

ECUADOR

EQUATOR

SOUTH AMERICA

SOUTH

ATLANTIC

INDIAN OCEAN

OCEAN

Scotia Sea

Weddell Sea

ANTARCTICA

PROLOGUE

It was one of the high points of my life. We had found the wreck of *Titanic,* lying more than two miles deep in the frigid North Atlantic waters, and it had set off a media frenzy. Top news anchors, late-night talk show hosts, and newspaper reporters all over the world wanted to talk to me, the man who had solved a 73-year-old mystery, the explorer who had found the most famous shipwreck in the world. And when I finally got home to my Cape Cod farmhouse, I realized that I hadn't had a chance to share the excitement with my parents.

Then the phone rang. It was my mom, calling from California. Ever since I was a rambunctious child, she had been my biggest booster. She had picked me up when I stumbled, encouraged my dreams, and taught me the discipline I needed to become an ocean scientist.

"Your dad, sister, and I have been watching," she said. "You've been on all the networks, and the phone has been ringing with friends and relatives calling us." I was leaning against the kitchen counter, soaking it all in. Then her tone shifted.

"But it's too bad," she said. "Now they are only going to remember you for that rusty old boat."

That rusty old boat. The phrase rings in my ears to this day. And you know what? She was right; mothers know best. When people hear the name Bob Ballard, they instantly think "the guy who found *Titanic*." Discovering *Titanic* was an experience of a lifetime and an accomplishment I'm proud of, to be sure. But there's so much more to my story.

Discovering new life-forms that rewrote our definition of life on Earth and possibly on other planets.

Finding other sunken vessels—*Lusitania, Bismarck, Yorktown, PT-109*—and answering questions that linger about their fate and those who perished with them.

Tracing ancient trade routes in the Mediterranean and the Black Sea, pulling up precious Roman, Greek, and Phoenician artifacts, and confirming hypotheses about a great flood in biblical times.

Developing robots that can roam the ocean floor 24/7, sending video back up to the surface, with views better than any we might have been able to get by diving in submersibles.

Offering "you are there" experiences to kids around the world through telepresence, so that even if they've never been on a boat or seen the open water, they can explore the ocean floor and hear scientists explain what they are seeing.

There are so many stories to tell, full of adventure, risk, thrills, and danger. There also is my own journey of self-discovery, as I learned to embrace failure as the greatest teacher and to set ambitious goals that help me achieve my dreams. Why join the crowd in climbing a 10,000-foot mountain, I like to say, when no one's tackling the 30,000-foot one nearby?

Perhaps you think of me just as the guy who found that rusty old boat. But let me tell you the rest of the story.

CHAPTER 1

FINDING NEMO

I was still a boy, bursting with energy and wanderlust, when I first discovered the tidal pools along the Southern California coast. Every day, when high tides recede, they leave behind a stranded menagerie of sea creatures in every niche and depression, only to be washed away at the next changing of the tides. It's like a kaleidoscope, and every 12 hours brings a new cast of characters. One time, we see sea urchins or a small octopus trying not to be seen; the next, a cluster of snails and small fish. Or maybe a crab that fights to the death when you try to pick it up, or a sea anemone waving seductively in the gently moving water.

I loved beachcombing—still do—wandering up and down the coast, picking up shells and the occasional glass float covered in barnacles. In grade school, I dreamed up a way to hook buried razor clams with a twisted coat hanger. I was always eager to see what was a little farther down the beach. Something magical might be just beyond the next point.

I got my first taste of the ocean, literally, when I was just two years old, and my father had taken me and my older brother,

Richard, to the beach. We gazed at the sea in wonderment, Dad said, like it was the greatest thing we'd ever seen, and before he could stop us, we both rushed headlong into the surf. Within seconds, waves flipped us head over heels, and I absorbed the first lesson the ocean taught me as Dad dragged us back to dry sand.

As a kid in Southern California in those days, you only went inside to eat or sleep, and I was on a tear from the moment I walked out the door. I was always a bundle of energy, and if Richard and I weren't at the beach, I was out in the yard, ignoring other restrictions. The house we lived in when I was a toddler had a backyard with a six-foot fence and an old chicken coop. One day my mother got a call from a local grocer saying I'd wandered into his store. She thought I was still in the backyard. I was marched home, spanked, and put back outside. Soon the grocer rang again, a little more irritated. I was back. Turns out, I had discovered a knot on the side of the chicken coop that I could use to leverage myself onto its roof and, from there, over the fence.

The next time I did that, though, I dropped smack into my mother's waiting arms.

I wouldn't quit, though, and another time she found me hanging from the fence by my suspenders. After that, she connected me to the clothesline with a leash. "Just like a puppy dog," I said plaintively. I was hyperkinetic, as they called it in those days. My mother was my biggest supporter, but even she got exasperated at times, trying to keep up with me. I remember her joking, "If you'd been born twins, I'd have drowned you both."

They say the sea calls to a person. Melville and Conrad wrote books about it. For me, maybe it was the ever changing creatures in those tide pools, or the sounds and smells of the surf that lured me. All I know is that my love affair with the sea began at a very early age, and it provided a hypnotic outlet for a boy who could barely sit still, a perpetual motion machine.

I was the middle of three children. Richard, my brother, was two years older, and Nancy Ann, my sister, was four years younger than me. Richard and I were fiercely competitive, something our father encouraged. Richard was off-the-charts brilliant, while I had trouble reading and felt confined in school. I tend to learn by seeing and doing more than by reading and studying. Then I fell under the spell of a mythical undersea explorer and landed an internship that gave me my first taste of life aboard a scientific research vessel. Each of these things lured me closer to the ocean and set me on my life's path before I was out of high school.

—

I WAS BORN IN KANSAS—and, given that state's landlubbing nature, I usually get a laugh when I tell people that's where all oceanographers come from. I was a war baby, arriving not quite seven months after the attack on Pearl Harbor, and my father was about to start a job as a supervisor at a Wichita Boeing plant that made B-29s. His next job took us to the Los Angeles suburbs when I was two, and after that we lived briefly in the Mojave Desert, where Dad worked with Chuck Yeager and other hotshot test pilots. I remember going with Richard into the desert to explore. We would sit inside the wrecks of crashed planes and pretend we were flying. We also used to go out at night in an area thick with tarantulas, hunting them with flashlights. It was risky, I guess, but even as a child I was comfortable taking chances.

Dad was a test flight engineer, riding along as pilots pushed the speed boundaries. It was dangerous work, so he hid the details from my mother. If she called while he was in the air, they'd just tell her he was "upstairs," and she just assumed there were no phones on the second floor.

When my mother finally learned what "upstairs" really meant, she insisted that Dad quit and move the family back to Kansas.

Reluctantly, he did, but soon we were back in California, and from the time I was six until I was nearly 11, we lived in Pacific Beach, the palm-lined part of San Diego just north of Mission Bay, where I found my tidal pools. And for the rest of my life, I would never live far from an ocean.

In many ways, ours was a classic Ozzie and Harriet family, just like the one in the 1950s TV show. My mother was even named Harriett. There also was an air of mystery about my father and a bit of tension along with love around our dinner table.

We never knew much about my father's early life, except that his father, a Kansas police officer, had been shot in the head and killed when Dad was five. His mother died when he was 12, and he was put on a train to go live with an aunt and her uncaring husband on a ranch in Montana. Other than telling us he grew up as a cowboy, my father didn't talk about his childhood. He dropped out of college after a year or so—a fact he concealed from us until we were in college—but he figured out how to make himself into an engineer, rising to chief engineer for North American Aviation on the guidance systems for Minuteman missiles.

He did a fine job for a kid who was basically tossed in the trash, and our family meant everything to him. He took us on excursions most weekends: camping, fishing, or just hitting the beach. Every summer, we'd head up to the High Sierra to fish for trout or to Monterey Bay to fish for mackerel and albacore. But the truth is, I didn't spend a lot of time with him. Dad took Richard to his office to meet the engineers, but he never took me. I think he saw in my brother what he had always wanted to be.

I'm sure I got some of my determination and tenaciousness from my dad, but I was a lot more like my mom. She was outgoing and fun-loving. We called her motormouth. She had been an honors student, but she never finished college. She married my father and stayed at home with her kids. She was my protector, my champion. Whenever I stumbled, she was right there, ready to take me into her arms. She brought the emotional quotient and humanity into the family.

I knew a lot more about her background. Her German grandfather had fought for the Union in the Civil War and then settled in Kansas. She came from a family with that basic midwestern common sense, and they taught me about discipline and setting goals. My grandmother was a devout Lutheran. When she heard I was studying evolution in school, she said, "I've lost my grandson to the devil." She was full of wise sayings. "Greatest is the person who plants a tree knowing he'll never sit in its shade," she would say to me. Thoughts like that encouraged me to think about things larger than myself. In our family, I was always more aligned with the women than the men, and I still identify strongly with the Kansan in me.

Once I got past the toddler stage, Mom gave me license to roam. "Just be home before dark," she would say, and away I'd go. When we lived in San Diego, I could bike to a fishing pier that attracted a wild assortment of croakers, surf perch, mackerel, and corbina, my favorite eating fish. Mom would drive me up to La Jolla Cove and leave me to spend the day fishing and swimming. Just to the north, I could make out the long pier at the Scripps Institution of Oceanography, the oldest and largest ocean research center in the United States.

Before long, with my mother's encouragement, I decided to go commercial with my fishing. I made a smelt trap out of chicken wire, and a friend, Johnny Binkley, and I would load it with bits of soggy bread. In less than an hour, our wagon would be full of flopping fresh smelt. I was a friendly kid and a natural salesman, and I loved selling the fish door-to-door on the way home, being careful to save enough for our families.

Dinner table dynamics were complicated. There sat Richard and my father, two serious peas in a pod, bookended by me and my mother, a pair of chatterboxes, with Nancy Ann in the middle. Even now, I believe that Richard was the smartest human being I ever met, and I've lived my life surrounded by smart people. It wasn't easy to follow him. It left me feeling like I had something to prove—a feeling that has motivated me my entire life. There's a difference between an A+ and a mix of A's and B's. I'd walk into a class a couple of years after Richard, and the teachers expected me to be like him. Some were hard on me when I wasn't. I received mostly B's and C's in citizenship—the euphemism back then for behavior—with comments from teachers that I needed to quit talking out of turn and show more self-control.

Even in grammar school, I couldn't keep up with the reading. How do you overcome being unable to read fast? You read for twice as long. Richard would be asleep, but I was still up, doing my homework. Writing was as hard as reading. When we would fight, Dad told us we had to put our grievances on paper. Richard could crank out a page or more almost instantly, while I had trouble writing a sentence. I hated it so much that I often settled with my brother "out of court" so I didn't have to go through the ordeal. My mom worked with me on my organizational skills, taming my scattershot brain. She always told me I had trouble reading and writing because I'd missed a phonics class during one of our moves. I accepted that. OK, fine, I told myself. I'll just work harder.

My sister had medical issues that deeply affected our family. She was born with a genetic mutation that gave her an unusually small lower jaw and, despite reconstruction, she could never speak. She wrote only in a rudimentary way, too, but she was smart enough to keep track of sports teams and the high-interest loans she made to me and Richard from her allowance. My parents took her to specialist after specialist. One drilled a hole in her head to relieve the pressure, real voodoo stuff. But the mix of things she could and couldn't do was so peculiar that doctors never found a case study to explain it. Finally, my dad just said forget it, we're not going to torture her with tests anymore. He never talked down to her, and our weekend excursions were always to places that Nancy could enjoy. It wasn't until late in her life that I was able, through DNA profiling, to figure out the cause of her disability. A transmutation at the moment of conception meant extra amino acids attached to her 15th chromosome and reprogrammed her body.

Despite disabilities, Nancy never got angry, and a smile rarely left her face. Seeing what she had to deal with helped temper my own insecurities. I'd look at her and think how small my problems were and how I was not going to waste a second of my life.

Looking for ways to distinguish myself in my father's eyes, I became the family clown, the prankster. Richard was more of an introvert, so I was a ham, playing Peter Pan in elementary school. Richard was not merely a Boy Scout, but an Eagle Scout, collecting merit badges and eating them like popcorn. I went wider and channeled my hyperactivity into sports, student government, and salesmanship. My mother would send me door-to-door, selling foam Santas and floating candles, hitting up my old smelt customers.

When I was 11, we moved to Downey, a Los Angeles suburb. I was still struggling with reading—I received a D on my first report card there—and my parents arranged for daily tutoring. I guess it helped a little. But I preferred going to the movies over reading books. I vividly remember, when I was 12, crowding into the theater for Disney's holiday blockbuster, a big-screen adaptation of Jules Verne's *Twenty Thousand Leagues Under the Sea*. It blew my mind.

In one scene, Kirk Douglas comes across Captain Nemo's submarine, *Nautilus,* seemingly abandoned in open water. He wanders around inside the weirdly luxurious sub and, through a portal, watches divers conducting an underwater funeral in a kelp forest. They were walking along the ocean floor, wearing these amazing diving suits. That's what did it for me.

As much time as I had spent around the water, it had never occurred to me that there was a vast, three-dimensional space beneath the ocean surface and that all sorts of amazing creatures lived there. I wanted to be Captain Nemo. I wanted to walk on the ocean floor. To my parents' credit, they never laughed when I said that.

My grades started to improve. I had a dream now, and I needed good grades to achieve it. Trim and heading toward six feet two inches, I also was a good athlete, playing football, basketball, and tennis. My social life revolved around clubs sponsored by the local YMCA with names like the Crusaders and the Visigoths.

Those were good years. Downey had one of the very first McDonald's. Burgers were 15 cents, fries a dime. Gas was around 30 cents a gallon. The American Century was in full swing, and young people in California were reaping the benefits. Dad bought a 32-foot wooden cabin cruiser and named it the *Nancy Ann,* and we headed out to sea pretty much every summer weekend.

When I was a junior, Richard went away to Berkeley. He got a full ride from Dad's company and majored in physics. But he suffered a serious health setback in college. His intestines became inflamed and had to be partly removed. It didn't hold him back from graduating with flying colors, but it turned out to be a precursor to problems that would plague him the rest of his life.

With Richard at college, Dad had more time to spend with me. It felt great. He did battle with my German teacher over a bad citizenship grade that threatened to keep me from earning life membership in the California Scholarship Federation, a big deal that would go down in my record and enable me to wear a white robe at graduation. Dad also arranged for a visit to Scripps, the oceanographic institute I'd seen in the distance when I was fishing in La Jolla.

—

THE RESEARCH FACILITIES AT SCRIPPS, which is now part of the University of California system, stretch out along the coast between a bluff and the foaming Pacific surf. My dad had met the director, Roger Revelle, during some of his military work, so we received a VIP tour. I told Dr. Revelle that I dreamed of being an oceanographer. He advised me to get an undergraduate degree in a technical subject—math, chemistry, geology, or physics—and then study oceanography at a graduate school like Scripps. I was so excited that we went back a second time to meet the dean of students, Norris Rakestraw. After we got home, I applied for a summer program at Scripps that was funded by the National Science Foundation, and Dr. Rakestraw sent me a letter inviting me to participate. That June, in the summer before my senior year in high school, I reported to Scripps as a junior trainee for a series of three excursions. I celebrated my 17th birthday at sea. Our initial work was part of a study looking at why sardines—immortalized in John Steinbeck's *Cannery Row*—had disappeared off the California coast.

My first official ocean cruise—of the 157 I have now made—lasted six days aboard a former Army cargo ship. To chart the currents, we dropped so-called drift bottles into the water, with messages inside offering rewards to whomever returned them. The motley crew members enjoyed springing tricks on newcomers, and I attracted my share. One time, they told me to take the larvae and other tiny creatures we'd captured in our net down into the stifling engine room and put them into a certain bottle. What they didn't say is that the bottle contained formaldehyde, a preservative that smells so strong it can make you vomit. How I managed to hold down my lunch, I cannot tell you, but I was on full alert after that.

My second cruise that summer—aboard R/V *Orca*, an old Navy tugboat—was much more adventurous. We were to be at sea for three weeks, working our way up the coast toward Oregon. I spent a lot of time harnessed into what was called the "hero's bucket," an open metal basket that hung from the side of the ship just above the water's surface. As the waves crashed over me, I attached a line of bottles to a wire that was lowered into the sea to collect water and measure temperatures at different depths.

We operated on an around-the-clock schedule, working four hours and then getting eight hours to sleep, eat, and rest as best we could. The *Orca* was small and constantly rolled around in the waves, and still, I couldn't believe how much fun I was having. I loved fishing for albacore, which can be an art form when the ship is steaming at full speed. I was so good at reeling them in that crew members stopped playing tricks on me.

The routine continued with little interruption until we stopped in Santa Barbara, where we were greeted by Robert Norris, a professor at the University of California campus there. He was a prominent figure who had earned his Ph.D. at Scripps under Francis Shepard, known as the godfather of marine geology. I told Dr. Norris about my interest in oceanography, and he urged me to consider studying marine geology at UC Santa Barbara.

Soon, we were heading up the coast again. The sun set over the ocean with an escort of billowy clouds, and that night a storm forced us to stop at a small port. I stayed on board, too young to drink, but the rest of the crew had a night of shore liberty. All tanked up and back aboard, the cook and the winch operator exchanged words, and a fight broke out between them. The cook lost his Coke-bottle glasses, stumbled into the galley, and came out waving a butcher knife at the first object he saw—which happened to be me. I jumped onto the mess table and hung from the pipes overhead until he was subdued.

We were several hundred miles out in open water when we ran into a more powerful storm than just about anyone on board had ever encountered, with gale-force winds and giant waves breaking over the bow. We pointed the ship into the waves and rode each swell up to the summit, only to plunge down into a trough and then work our way back up again. The waves threw the winch operator out of his bunk and broke his hip. All we could do was give him morphine for the pain. We couldn't head back to shore— the waves would have rolled us over—so we had no choice but to wait, sleeplessly, for help. One day passed, then two.

Finally, the Coast Guard dispatched a cutter to help us. As I watched through binoculars as it approached, a giant rogue wave hit its starboard side with such force it almost knocked the ship under. "Oh my God, look . . ." was all I could get out before the same wave hit us square on the nose with such force that it smashed the windows near where I was standing on the bridge, blew out several portholes on the mess deck, and injured another crew member. We ran around like madmen, stuffing life jackets into every opening we could find, then building a cofferdam of mattresses around an opening to the engine room. At last there was a break in the weather, so we could turn back home, but the damage to the ship was so severe that the cruise was terminated.

It was amazing, and humbling, to see the force of our Earth, but I don't remember being scared. Going through that storm taught me that there were forces I had to respect. I couldn't just bull over everything with my energy.

When I got back home after that adventure, I was given a hero's welcome and a big hug from Mom. At dinner, the evening was moving along very nicely until I asked someone without thinking to "pass the fucking butter." The dining room went silent. My mother finally chirped, "Well, my son is clearly becoming a sea-soned sailor."

—

THE FINAL SCRIPPS EXCURSION was led by Carl Hubbs, who epitomized the absentminded professor. He had set up a field station to study sea life at Punta Banda, on Mexico's Baja Peninsula. I hopped into his pickup, and we headed toward the border. As we approached a military checkpoint, Dr. Hubbs turned to me. "You don't have a visa, do you?" I didn't even know what that was. I'd crossed into Mexico before with just my driver's license—but to work there, you needed a visa, I was only now learning. Dr. Hubbs yanked the wheel and charged into a farmer's field, mowing down a row of crops and circling around the checkpoint. I thought my guts were going to fly out of my mouth. And I'm happy to report that we did not end up in a Mexican jail.

The field station was primitive, just a pair of Army tents on a rocky bluff overlooking a cluster of beach shacks. The tents were like ovens, so we slept outside. The work was constant and grueling. I was tramping all over the place, measuring wind velocity, taking water temperatures. Hoping for a respite from scalding canned food, I decided to go fishing—but when I brought back my first catch, Dr. Hubbs just grabbed it and put it in formaldehyde.

Soon Dr. Hubbs departed for another study in Alaska, leaving me in the hands of a pair of new Scripps Ph.D.'s. That night, they took me into town for what they said was going to be a real dinner. Soon, I found myself in the front row at a Mexican strip club. I'm sure it was prearranged, but one topless dancer shimmied over to me. I was terrified. I mean, I was just some kid getting his first taste of the real world. My eyes never left her face, no matter how much she jiggled, while the other two guys stuffed money into her G-string as fast as they could.

The two of them taught me something else, too. I said I was considering marine biology as a major, just like them. They tossed a bucket of cold water on that idea. "Major in anything but marine biology," they told me. "There are no jobs to be had."

Those words rattled around in my head after they returned to San Diego, leaving me alone at the camp. I was guzzling water 24 hours a day in that awful heat, and I found an open barrel marked "spring water." After passing out, I woke up violently ill, purging in every direction. At times, I was hallucinating. I was afraid I was going to die, then afraid I wasn't. It turned out that the water had been meant for the jeep's radiator, not for humans. I also managed to get a cactus spine stuck in the side of my foot, and it got infected.

Finally, Dr. Hubbs returned. He dug the spine out of my foot and drove me home. When I got there, I said, "Dad, I definitely don't want to be a marine biologist."

But I needed to make some sort of decision about a school and a major. College was approaching fast. Despite all that had happened, I knew the future I wanted. I wanted to be out there, in the field, exploring around the next point, mixing scientific achievement with physical exertion.

I wanted to be Captain Nemo. Maybe I couldn't be in the movies, but I intended to walk on the bottom of the ocean.

CHAPTER 2

SWISS ARMY KNIFE

I was getting ready to head off for college, and we had just moved to Long Beach, into one of those classic California split-levels they used in TV shows to show the good life, palm trees and all. I remember sitting at the kitchen table with the University of California at Santa Barbara catalog in front of me. It was an inch thick.

I flipped eagerly from page to page, major to major, class to class. College looked like the greatest smorgasbord in the world, and I did not want to miss a single bite.

I knew I wanted to be an oceanographer. That was a given. But oceanography was a graduate program. I needed to get a solid foundation in one of the sciences. I was pretty confident that, with the connections I'd made and the cruises I'd been on, I would end up getting my Ph.D. at Scripps, and I was looking for a major that would set me up nicely for that. Dr. Revelle had suggested I study math, chemistry, physics, or geology. I saw something in the catalog called a physical science degree. It was ambitious. Students majored in two of those four fields and minored in the other two, and the degree required five years to complete. Maybe I should have taken pause, given my problems with reading and test taking. I'd scored only a 444 out of 800 on the verbal SAT and 538 on the math. But I still thought I could get the program down to four and a half years if I went to summer school. It was a souped-up, 450-horsepower version of what Dr. Revelle had suggested, sure to impress the Scripps gatekeepers—not to mention my father and brother, who was breezing through physics at Berkeley and heading for a Ph.D.

So off I went to UC Santa Barbara, with its campus sitting along beautiful cliffs above the Pacific and wrapping around its own lagoon. Pledging a fraternity was a no-brainer since I'd liked the YMCA clubs so much. Two years of Reserve Officer Training Corp (ROTC) was mandatory for men, and even though our campus was on the coast, the only service there was the Army, not the Navy. I eagerly joined, because I had grown up in military towns and felt a responsibility to serve. I had turned down a basketball scholarship because my dad was worried it would interfere with my studies, but I still played on the freshman team. I had money from my after-school and summer jobs, and my dad had me sit at the kitchen table and draw up a budget. I got so into running the numbers that I asked Mom how many sheets were on a roll of toilet paper so I could figure out how many rolls I'd need. Dad said I didn't need to be quite that detailed.

I managed an A in chemistry my first semester but struggled to eke out C's in math and German III. I had to have an impacted wisdom tooth removed while I was studying for the German final. Half under, the surgeon later told me, I kept answering his questions in German.

I decided to major in geology and chemistry—geology because Dr. Norris was a marine geologist, and chemistry because I had gotten an A in the first class. I had to take all the core courses in physics and math, too.

I was able to hold my own in most classes with a visual trick I'd taught myself in high school: photographing notes in my mind and retrieving the pictures of them during tests. I was shocked, however, when I got D's in differential equations and organic chemistry. Some of the questions involved word problems, and I had trouble conceptualizing them. The words went into my head and just broke apart. I would read them, read them, and reread them, but it didn't click.

The next summer, I took some courses and also worked for the ocean systems group at my dad's company, North American Aviation. With President John F. Kennedy's pledge to put a man on the moon, companies like North American, which is now part of Boeing, had expanded into spacecraft and were starting to look at the deep sea as the next frontier worth exploring. My job there that summer was to put together hundreds of three-by-five-inch cards on the history of deep-diving submersibles.

In my junior year, I was elected class president, and I managed to win over a classic California beauty, whip smart, named Lana Rose. We went to fraternity parties and ROTC balls, and just being known as the guy who won her boosted my status. Not long after we started dating, I was so excited we drove home together and got there so late that we had to roust my parents out of bed to meet her. Even at that hour, she charmed everyone.

It was the early 1960s, and national politics was heating up on most college campuses. I had grown up in an Eisenhower Republican household, and Santa Barbara was very conservative. You didn't hear much talk about JFK's Camelot. We knew that college students elsewhere had joined the Freedom Riders to push for civil rights in the South. But the larger rumblings of the social unrest that would soon unsettle American campuses had yet to appear. The local chapter of the John Birch Society, a far-right anticommunist organization, was still a force. Its members objected to a campus appearance by Margaret Mead, the anthropologist whose studies of primitive cultures had rattled some right-wingers. I got involved early in that fight in support of free speech, and she eventually got to talk.

American involvement in Vietnam was still in its early stages. The Pentagon had sent advisers to help South Vietnam counter aggression from the Vietcong in the north, but there was no doubt in the minds of the officers who ran our ROTC program that the United States would end up sending troops there. This really began to snap into focus for me at the start of my junior year, when I had to decide whether to become an Army officer or drop out of ROTC and risk being drafted after graduation.

If duty was likely to call, did I want to answer it with an officer's gold bar or the insignia of a private? And perhaps more important, did I have a taste and a talent for leading?

I wanted to find out.

Quite frankly, I'd enjoyed ROTC. I'd loved wearing a uniform in sports and in the Boy Scouts—I associated it with good qualities like discipline and teamwork. In ROTC, we wore our khakis or winter greens every Thursday before we drilled, meaning we spent Wednesday evenings perfecting spit shines on our shoes. I also had a talent for marksmanship. My father had given me his .22 rifle when I was a boy, and I had used it while visiting my grandparents in the hills near San Diego, where the walnut farmers were pestered by nut-hoarding squirrels. They paid me 25 cents a tail, and I learned to make every shot count, since I was paying for the bullets. Each year, I rose in the ROTC ranks. I was chosen the top cadet as a sophomore, and after deciding to stick with it and become an officer, I was named top cadet again as a junior.

ROTC gave me a sense of how visually oriented I am. Air Force officers visited our unit with a flight simulator and asked us to alert them when we saw a tank, thousands of feet down on the ground. As soon as they turned it on, I said, "See it." One of them said, "God dammit, you're right." I did that over and over, and they asked if I wanted to join the Air Force.

The bigger test came the summer after my junior year, when I spent six weeks at Fort Lewis, south of Tacoma, Washington, undergoing what the Army called leadership reaction training. Basically, the instructors would try to throw you off balance under stressful circumstances to see how you responded—and they were not messing around. An instructor would sometimes walk up and say, "You're dead," even though you had done nothing wrong. You were just the one the imaginary mortar shell happened to hit. One drill sergeant was killed when a cadet inadvertently discharged his M1 rifle inches away from the sergeant's kidney. We were using blanks, but even blanks can kill you at such close range.

We were divided into teams, usually with four cadets each. Two teams made a squad, and four squads made a platoon. Each night, you were assigned a rank for the next day. One of us would lead a team and be given a mission, and then the instructors would change the mission halfway through—or "kill" the team leader— just to see how we adapted. They would be yelling at us, and you learned pretty fast if you could cut it. At some point, everyone had to ask himself, *Am I a follower or a leader?*

One mission involved capturing an enemy soldier as your squad was passing through wooded terrain to get to the front lines. There you found a minefield to your left, a minefield to your right, and a barrier of razor-sharp concertina wire straight ahead, like a Slinky lying on its side. The first cadet got through the concertina wire, and then, while the second was crossing, the enemy opened fire. Now what? Your team is cut in two. The test was to see whether the leader would order the man caught in the wire to lie down and allow the rest of the squad to step on his back to cross, pressing him into the lacerating wire. The idea was to drive the enemy back, reunite your squad, and then extricate the man from the wire and attend to his wounds before proceeding with the mission.

One day, as we formed for breakfast, the drill sergeant checked our faces to see if we had what he called "a good Army shave." We didn't, and he sent us back until we had shaved so close we had razor nicks all over our faces. For the first drill, we assembled in a field of tents. We were told to put on gas masks and go into one of the tents, where a metal trash can was pumping out tear gas. We were ordered to remove our masks and give our name, rank, and serial number. As soon as you took off your mask, the razor cuts started to burn, like someone had thrown acid on your face. If you were lucky, you got your name and rank right. But everyone lost control and began to cough violently before spitting out our serial numbers.

Then—with snow-covered Mount Rainier in the background, just as rain began to fall—we assembled on metal bleachers, and the sergeant talked about how to survive a nerve gas attack. Each of us was given a small tube with a needle sticking out of it, to simulate an antidote kit. Stand up, we were told, drop your fatigues to your ankles, and sit back down.

"Now place the tube firmly in your hand with the needle between your two middle fingers," the sergeant told us. "What I am going to say is, 'one, two, hit it,' and on 'hit it,' drive the needle into your thigh a little off to the side, to miss the bone."

We waited tensely for the command, pants around our ankles, needle in hand. "One, two, go," the sergeant yelled, and we drove the needles into our thighs.

The sergeant scowled at us. "I didn't say 'hit it,'" he said. "Now pull it out; we're going to do it again."

We yanked out the needles, sending trickles of blood mixed with rain running down our legs. I could hear helmets crashing down on the metal bleachers as some cadets passed out. The rest of us listened more carefully the next time.

"Now, I am going to come around and hit the top of the tube, to make sure the needle is all the way in," he said. After this lovely piece of instruction and his painful taps, we were ordered to reassemble in platoon formation.

As the brigade commander that day, I had to lead the platoons across the field. Then, all of a sudden, canisters of gas were lobbed at us, and most of the men dropped to the ground, retching violently. I was out in front, not hit by the gas, but the sergeant let me have it as I exited the field. "Cadet Ballard? Where are your men? Return to the field and don't come back until all of your men are accounted for!" Soon I was on the ground, vomiting with my fellow cadets.

Over those six weeks, we fought with bayonets and crawled under barbed wire with live rounds flashing over our heads. I became an expert rifleman, scoring 94 out of 100, and came in sixth out of 1,400 cadets in physical fitness. I also learned lessons that were crucial to me later in perilous moments at sea. I learned that you can't know the true measure of people—who's going to run and who's going to stand—until the bullets start flying. As the drill sergeant screamed at us, some people buckled, some freaked out, and some quit. But in those tensest moments, the whole world slowed down for me. I became extremely calm, looking and thinking and trying to figure it out. I learned that leadership involves rallying your people when things go bad.

But I also was taught that if someone runs, I should turn and shoot him, or else I might lose everyone else. Would I ever have done it in the field? Probably. Would I have been able to live with myself afterward? Probably not.

—

I WAS RIDING HIGH when I got back from Fort Lewis. I took some late-summer classes at UCLA. I met Arthur Ashe, the future

Wimbledon champ, who had just arrived at UCLA on a tennis scholarship, and I even beat him in a tennis tournament. Back in Santa Barbara for my senior year, life was good. I became the deputy brigade commander of our ROTC unit. If I needed to, I could deliver a gruff monologue like General Patton or an inspirational call to arms like Henry V's "band of brothers" speech on St. Crispin's Day, which we had studied in military class. Leadership was somehow part of my DNA. I was enjoying my advanced geology and chemistry classes. To the envy of my fraternity brothers, I got engaged to Lana.

But then, late the next spring, things began to sour. Out of nowhere, Lana dumped me. I was numb and embarrassed. I spent a lot of that summer scuba diving, often along an isolated stretch of the Big Sur coast, south of the Monterey Peninsula. I told my father that my diving was all about the pursuit of science. But he could probably tell it was to keep my mind off Lana—and maybe off my future, too.

I had my sights set on Scripps for graduate school. I'd been telling myself that my smattering of D's would not stop them from accepting me into the doctoral program, because I'd interned there in high school and kept up with some of the officials. Scripps did research for the military, so I thought they'd appreciate my time at Fort Lewis and my ROTC designation as a Distinguished Military Graduate.

But lurking in the back of my mind was a question mark attached to that dream. My overall GPA was hovering around 3.0 as I entered my ninth and final semester at UC Santa Barbara. I knew it was low for a Scripps applicant, but I hoped my interdisciplinary major might make up for it.

As part of my application, I drove down to La Jolla for an admissions interview with Fred Noel Spiess, director of the Marine Physical Laboratory at Scripps. The son of a Navy sonar man during World War I, Dr. Spiess had survived 13 submarine patrols during World War II, earning both a bronze and a silver star for gallantry in combat. He had a master's in engineering from Harvard and a Ph.D. in physics from Berkeley, and he was a formidable, highly accomplished oceanographer.

I had never met Dr. Spiess, and I went into my interview full of enthusiasm. But he was cold and clinical. It was not a long interview, and I remember walking away thinking, *Well, I don't know how that went.*

I found out soon enough, when a letter arrived from Dr. Spiess. I'd been rejected.

I had known this was possible. But I had convinced myself it wouldn't happen. Now it felt like all my dreams were crashing. I was in despair. I was doubting myself. *I'm just not cut out for oceanography,* I thought.

What made it worse was that I felt I could have done it differently. I had chosen an outrageously difficult major. I had gone too wide. Athletics, student government, Greek life, ROTC. I was like a Swiss Army knife, a master of nothing but pretty good at a lot. I should have locked myself away and studied more.

I also found out later that the recommendation letters I'd been counting on had contained stinging rebukes. My adviser, Dr. Norris, called me "personable" but a little too absorbed in the "social whirl." Another professor praised my "dynamic outgoing personality" and noted I was exceedingly versatile, with "unusual ability," but "inclined to do only as much as it takes to earn an honors grade and no more."

I probably told my brother about the rejection, but by this point in our lives we were not close enough that I would have sought his advice. Besides, he was married and working on his Ph.D. in particle physics at Berkeley, and his successes made me feel even more a failure.

The fire was burning pretty low. Everyone seemed to be telling me to chuck it, become a salesman or go into business or something. I knew that I had to pick myself up, rally like I'd learned to do with my troops, and keep moving forward in the face of hostile fire. I remembered watching Mexican fighters on television and how those smaller fighters would basically tell their opponents, "I'm going to let you break your hand on my face." They'd get knocked down and then get right back up. I admire that quality in a person.

So I got back up, too.

—

MY PASSION FOR THE SEA kicked back in, and I embarked on a mad scramble for fresh options. In my single-minded focus on Scripps, I had not researched other graduate programs, and graduating in the middle of a school year didn't help. I had lost my fiancée, I had lost my dream, and I wanted to go to a place where I didn't know anyone. I discovered that the University of Hawaii at Manoa, in Honolulu, offered a master's degree in oceanography with plans to expand to a Ph.D. program soon. It wasn't Scripps, but it could get me going again.

I had to rush to Honolulu so quickly, I missed graduation at Santa Barbara. I asked the Army for a delay in active duty and began looking for a part-time job. Dr. Norris may not have written the strongest letter of support for me, but he did tell me about his brother, Ken Norris, a UCLA professor who did summer research on whales and dolphins at the Oceanic Institute, east of Honolulu. The institute was connected to Sea Life Park, an aquarium that offered dolphin and whale shows. I zipped out there on a rented moped and soon had two jobs: training dolphins and whales for the tourist shows and helping Dr. Norris with his research when summer rolled around.

Sunshine, tropical breezes, and frolicking with dolphins turned out to be a wonderful way to restore my spirits. Dolphins are social creatures. They love to have you rub their skin. It was 1965, and in those days at a place like Sea Life Park, you didn't think twice about training them to do all sorts of amusing things. In Whalers Cove, Hawaiian girls paddled canoes and swam in the tanks with them. In the Glass Porpoise Theatre, we demonstrated how we trained the dolphins and how intelligent they were. They didn't just swim around, jump in the air, or do hula dances with their tails. We found ways to take their abilities to another level.

I trained one of my favorites—Keiki, an Atlantic bottlenose dolphin—to place swimming pool floats in his "piggy bank," a bicycle basket suspended just underwater. I also trained him to bring floats to me. I would give him one fish for a white float, five for a red float, and 10 for a blue float. He quickly figured that out and brought me his blue floats first. He earned floats during the show, and he would store them in his piggy bank until he could trade them for fish.

We did four shows a day, and I learned how to read an audience and improvise to keep them entertained. I'd hit Waikiki at night with the other trainers. One was the daughter of a local radio personality, and thanks to her, we'd be treated like mini-celebrities at our favorite restaurants. Then we'd head off to hear Don Ho, the Hawaiian singer soon known nationwide for his song "Tiny Bubbles." The key was to arrive near the end of his last show so we only had to order one drink instead of the two-drink minimum.

I started dating again. A lovely brunette who lived across the street, Marjorie Hargas, came over to introduce herself one day. She'd spent part of her childhood in Montana, like my dad, and had set out to explore the world by herself, with her first stop in Hawaii. I thought that was gutsy. We tooled around together in my prized 1958 Chevy Impala convertible. That had been the debut year for Impalas. I'd always wanted one, and I found one at a great price there in Honolulu. It was a purplish color, and we called it "The Grape."

When I went home for Christmas that winter and told Mom that I had left Margie crying at the Honolulu airport, my parents sent her a plane ticket to come join us. People got married younger back then. My parents didn't pressure me, but I knew they'd be happy if I did. Richard had been married for three years, and he and his wife already had a son. Proposing seemed the right thing to do. Margie and I were having fun together, and to me, the model of marriage was the one I'd grown up with. I gave her a ring, and when we returned to Hawaii, we were engaged.

—

MY CLASSES WERE GOING WELL, and my work for Dr. Norris was fascinating. We studied how fast dolphins can swim and how they can dive to great depths without suffering the air embolisms that afflict other mammals, including humans.

I did not know that the institute was providing dolphin trainers to a highly classified Navy research project until I was approached one day about training dolphins to kill enemy divers in Vietnam. Turns out the Vietcong were sending swimmers into Cam Ranh Bay at night to place explosives on American supply ships. The bay was full of floating debris that made it hard to pick out the swimmers from other sonar targets. The idea was that the dolphins could detect the divers and let their trainers know. The trainers would attach a spear with a gas canister on the dolphin's nose. The creature would swim back and shove the spear into the diver's chest, inflating his dead body so it floated.

It didn't feel right to put the animals in that position, so I politely declined to get involved.

Not long after that, I had another disturbing encounter involving the war in Vietnam. One of my ROTC classmates was on R&R from his Army Intelligence job, and I met him and one of his buddies for drinks at Fort DeRussy on Waikiki Beach. I asked a couple of times what they were doing in Vietnam, and my friend finally said, "You don't want to know." I said I really did, because I would likely be sent there.

After a couple more drinks, he opened up. "I'll tell you what you will be doing," he said. "You will take three Vietcong who were just captured on patrol up in a helicopter, throw two of them out, and then interrogate the third."

I didn't sleep that night. There was no way I was going to do that. The next morning, I went to the recruiting office at Pearl Harbor and filled out a form requesting a transfer from the Army to the Navy, which a distinguished military graduate could do. Then Margie and I flew to California to get married—and, I thought, to live. My boss from the internship at my dad's company—Andrew Rechnitzer, a legend in the deep-sea world—had offered me a full-time job back in Long Beach, helping to design a deep submersible for the oil industry. The company also was willing to pay for me to get a Ph.D. at the University of Southern California. It seemed like the perfect chance to get back on a faster track.

We were still settling in, a few months later, when there came the proverbial knock on the door. It was a Navy detailer, telling me my transfer had been approved. I was soon assigned to the Office of Naval Research's Boston location as the liaison officer with research scientists on several campuses, including Scripps's biggest rival, the Woods Hole Oceanographic Institution on Cape Cod.

So Margie and I climbed into a VW Beetle we'd just bought, taped a $1,000 cashier's check under the dashboard, and began our journey from the Pacific to the Atlantic coast.

CHAPTER 3

BECOMING A SCIENTIST

A s Margie and I headed east, our horizons expanded with each mile. We made our way through deserts and tumbleweeds for 12 hours to El Paso, then through another nine hours of parched landscape to Dallas. We pulled over to watch a huge flock of Canada geese heading north to Yankee country, just like us. It seems improbable now, looking back on a life spent exploring the planet, but at that time I had never even been east of the Mississippi River. We crossed it at Vicksburg and toured the battlefield where Grant starved the Confederates out of a fort on the bluffs. I walked the terrain, imagining the battle unfolding. After that, Atlanta—and then Washington, D.C., for my magical first view of the nation's monuments. Finally, up to Boston and our first taste of New England: a chaotic March snowstorm. I woke up the next morning, put on my crisp dress blues, ready for my first day at work—and immediately had to change back into jeans to dig the VW out of the snow.

It was a difficult introduction to the Northeast for a beach-comber from Southern California, and my first days at the Office of Naval Research weren't much better. I had been training to be a warrior in the Army, then I got a great job designing submersibles and working on my Ph.D. Being activated by the Navy had meant a sharp detour with a huge pay cut. But here I was. I figured I'd do my duty and then head back to California.

My work involved monitoring grants to research institutions—just pushing around a bunch of papers, really. I was the liaison officer to several schools—MIT, the University of Rhode Island, and the University of New Hampshire among them. By far my largest grant recipient was the Woods Hole Oceanographic Institution, about an hour-and-a-half drive south of Boston.

I knew very little about Woods Hole. I'd dreamed for so long of joining Scripps that I hadn't really focused on its East Coast rival, which also was home to a variety of research scientists studying the oceans. WHOI, as they called it, was located in a little New England village, quaint cottages with slate-shingle roofs, while Scripps was classic Southern California, rolling surf and all. In oceanographic circles, it's a little like the Yankees versus the Red Sox. And it didn't take long before I realized that I should spend a whole lot of time there.

There I was amid all the L.L. Bean apparel in my blue winter uniform. I wandered across the Water Street drawbridge over a narrow inlet connecting Nantucket Sound with Eel Pond, a small lagoon dominated by the Captain Kidd bar and seafood shack. And then, as I shivered on the bridge, I noticed something out on the pond.

Trim, white, 21 feet long, there it was: *Alvin*, a submersible capable of carrying three people down to explore the ocean to a depth of 6,000 feet. Owned by the Navy but operated by Woods Hole, it was on the pond undergoing some sort of testing. I was mesmerized by the sight of it and my sense of what it could do.

Alvin was the only submersible at any American oceanographic institution. Scripps didn't even have one. Up until then, explorations of the deep-sea floor were conducted from the ocean surface, using sonar and other remote detection techniques. I could see that *Alvin* would make it possible for scientists to go underwater themselves and maneuver around to explore with their own eyes. That made it my kind of sea craft. Already, it had proven its worth by locating a hydrogen bomb that had fallen from an Air Force bomber during a midair collision and had sunk to the bottom of the Mediterranean. The thought of working in a manned submarine that relied on direct visual observation was exciting—just the sort of science I wanted to practice.

That first day at Woods Hole, I made my way through a gauntlet of introductions. It was like meeting the gods. First came the director of Woods Hole, Paul Fye. When I bragged about going to sea on Scripps vessels, he hit me with a gimlet eye and said, "Well, it's nice to see you have finally worked your way up to the greatest oceanographic institution in the world."

After that I met Kenneth O. Emery, known as K. O., an expert on continental shelves and one of the most celebrated students of Francis Shepard, the famous marine geologist. I also made a point of meeting Bill Rainnie, a former Navy submarine officer who ran Woods Hole's Deep Submergence Group. He was *Alvin*'s gatekeeper.

Over the next two years, they all welcomed me into the Woods Hole circle. Dr. Emery even invited me to go out on Woods Hole research vessels as part of the science team, and my Navy boss said fine. It felt like I was back in graduate school. My dreams weren't on hold anymore.

I started going full throttle, total focus. *Alvin*'s crew chief—George Broderson, or Brody for short—nicknamed me the "White

Tornado" because of the way I sailed through Woods Hole in my summer white uniform, letting out all my pent-up energy.

Call it luck. Call it whatever. Somehow, I had ended up in the perfect place, and I found myself being embraced like a member of the family.

—

THROUGH A CONTACT AT MY NAVY OFFICE, I started hanging out with the Boston Sea Rovers, the most prestigious dive club on the planet. They were a hardy, eclectic group of about 20 men who didn't mind freezing in the North Atlantic. We'd suit up and jump into the cold, dark water each weekend during dive season to search for lobsters, coming up with a supermarket full of food. One of my favorite guys in the group, Joe Hohmann, was so strong that he could carry two big air tanks on his back. He'd just stick his hand in a hole and let the lobsters bite him, and then he'd pull up these 20-pound suckers. At the end of the day, we'd all go to a deserted island, fill a trash can with moss, and cook up our catch. After each feast, we'd jump into the water to wash off the butter running down our wet suits.

For our annual weekend-long clinic, we'd kick off Friday night with a cocktail party at a grand place we called the Castle. Saturdays there were prominent speakers from the underwater world, like Jacques Piccard, who had taken the Navy bathyscaphe *Trieste* 38,500 feet down into Challenger Deep, the deepest part of the ocean. Saturday nights, we'd watch films that explored the oceans. Once, several members presented Frank Scalli, the master of ceremonies, with a huge lobster they walked across the stage like a dog on a leash, filling the room with raucous laughter. Each year's weekend wrapped up with a lobster bake at Frank's home. He was a scuba diver who represented U.S. Divers, a company owned by Jacques Cousteau, who was introducing scuba and underwater exploration to the world through television series like *The Undersea World of Jacques Cousteau.*

It's hard to express how much the Sea Rovers would influence my life. I watched how the best speakers drew us into their presentations with colorful underwater images. I met some of the best underwater photographers and filmmakers—Luis Marden and Bates Littlehales, for example, who worked for the National Geographic Society. I also met Melville Bell Grosvenor, the Society's president and editor of *National Geographic* magazine from 1957 to 1967. He was the grandson of Alexander Graham Bell, inventor of the telephone and first president of the National Geographic Society. And at Sea Rovers meetings, I met the most famous underwater personalities of the time: Al Giddings, the cinematographer who would film *The Abyss* and *Titanic,* among many others; Rod and Valerie Taylor, known for films that got up close with sharks; Harold Edgerton, an MIT engineer who worked with Cousteau and others on underwater technologies; Eugenie Clark, an ichthyologist and scuba pioneer nicknamed "The Shark Lady"; Paul Tzimoulis, early advocate of scuba and publisher of *Skin Diver* magazine; Joe MacInnis, known for intrepid polar dives; Stan Waterman, pioneer diver and creator of underwater films and television specials; and Peter Gimbel, who dove to and photographed *Andrea Doria*. Jacques Cousteau and his son Philippe often joined us as well.

I listened as some of the Rovers described how they loved to pull on their scuba gear and dive on the wrecks of small ships and pleasure craft. I got really caught up in that idea. I must have wanted to up the ante, because Joe Hohmann remembers that even back then, in the late 1960s, I talked about wanting to find *Titanic. Gee, I don't know what this guy is smoking or drinking,* Joe remembers thinking. He viewed my riffs on searching for *Titanic,* he says today, as just "a pipe dream."

—

THEN, IN 1969, the Navy upended my life again.

The Vietnam War was draining the Pentagon budget, so the Navy needed to make cuts. Junior research officers like me were given a choice: Commit to a full-time career in the Navy or get out. I wanted to get my Ph.D., but I also needed a job.

I will never forget how Dr. Emery and Bill Rainnie supported me at a time when I really needed it. Dr. Emery suggested I enroll in Woods Hole's Ph.D. program. But when other officials decided I couldn't enroll in a degree program while I also worked there, Dr. Emery bailed me out again, arranging for me to commute to the University of Rhode Island's Graduate School of Oceanography to finish my doctorate. Meanwhile, Rainnie said not to worry about a job—I was such a great talker that he'd hire me to promote *Alvin* in the ocean science community. *If only there was another sunken H-bomb to search for, we'd get all the attention we need,* I remember Bill saying—and I told him that if only *Alvin* could find *Titanic*, everyone would want to use it.

So I found myself out of the Navy and back in oceanographic research. The main thrust of my research was a geologic survey of the Gulf of Maine's seabed. The concept of plate tectonics was just gaining traction: the theory that the planet's surface consists of large crustal plates that bump into, grind against, or pull apart from one another in a massive synchronized ballet. Plates may move away from one another, creating new crust; or they may move toward each other, with old plates diving back inside Earth; or, as we now see in California's San Andreas Fault, they may grind past one another and generate earthquakes. This picture of the dynamic changes in Earth's crust occurring over eons is widely accepted now, but in those days, there were many skeptics. The theory made sense to me, though, and I felt that my research— which would focus on how the North American and Eurasian plates pulled apart beneath the Gulf of Maine, creating the North Atlantic—could help make the case.

Becoming a scientist means lots of reading—hundreds and hundreds of scholarly articles to catch up on the background and figure out what to focus on when I went out into the Gulf of Maine. Reading still did not come easily to me, though. When I took the Graduate Record Exam, I had scored in the 95th percentile in math, but my verbal score put me in the 41st percentile, 460 out of 800. I was aware of how much noise distracts me, too, so as I got into my research in the spring of 1970, I needed to find a place of silence where I could focus. Thank God for libraries.

Margie and I were living in a small house. Our son Todd was a year and a half, and another son, Douglas, was on the way. We needed more space, but we couldn't afford much. I spotted a 16-room farmhouse in nearby Hatchville, near Falmouth, Massachusetts—a wreck of a place that had been built in the 1600s and was one step from the bulldozer. It cost only $20,000, and I put down $4,000 from the severance pay I'd gotten from the Navy. I told myself it showed promise and figured I could fix it up. But when my mother finally saw it, all she could muster was, "You've gotta be kidding."

I also needed an outlet to burn off my excess energy, and I figured that rebuilding the house, room by room, would do the trick. I did all the plumbing, wiring, drywalling. I knocked down walls and busted up an old foundation from what was once the milking barn. I planted fruit trees to provide inexpensive food. Through it all, I developed a very disciplined ritual. I'd go up to the study I had built over the kitchen and read academic articles for 45 minutes. Then I'd reward myself with 15 minutes of work on the house, maybe smashing something with a sledgehammer. The key was to strictly limit my break time and get back to reading. I liked this pattern of mental, physical, mental, physical—which I maintain to this day.

I was working so hard just trying to make ends meet that I had little time for a social life. I didn't get to dive with my Sea Rover buddies much once we moved to Hatchville. The only friends I had were my work colleagues. It didn't help that Woods Hole was a small, insular community dominated by academics. Even the spouses tended to be highly educated. Margie had not gone to college, and I could tell that the intellectual atmosphere intimidated her. She didn't feel comfortable socializing with those couples.

I had never been around so many professional women myself. My mother, smart as she was, had been a housewife and stayed home with the kids, as most mothers did in those days. I began to realize that my brother and I had both opted to marry housewives like our mother. Margie's discomfort with the Woods Hole crowd was a little disconcerting, but I had so much else to do, I didn't stop to think about it. And Margie was making friends on her own with some of the local residents.

—

MY PH.D. RESEARCH took me out on surface vessels into the Gulf of Maine, a basin created as the continental plates pulled apart. My goal was to create a 3-D image of this now buried terrain, collecting data to document the theory of plate tectonics. I blasted a compressed air gun every 20 seconds to penetrate the seabed, creating images of the hidden structures beneath it as the shock waves bounced back to the surface. The constant discharges were annoying, especially for anyone trying to sleep. One crew member lost it and came after me with a knife. Who thought science could be such a violent pursuit?

I also was applying for federal grants to equip *Alvin* with more tools. The little white sub was in dry dock when I started my research. It had flooded and sunk when cables broke during a launch, and it was being refurbished.

Land geologists jump into their Jeeps to climb the sides of volcanoes and zip out onto the plains. I'm a field geologist, too, just an undersea one, and I wanted to convert *Alvin* into my own jeep—something I could jump into and drive along the ocean bottom, banging on rocks and retrieving samples that could help sort out the geology of the seabed. I wanted to add a diamond-bit drill to extract rock specimens, and I needed acoustic transponders that I could drop into the water to track *Alvin*'s movements more precisely. Improvements like these would provide better data and allow other scientists to verify *Alvin*'s discoveries.

Sure enough, with new equipment and a revitalized *Alvin*, I started getting interesting results. Once I had brought up rocks from the floor in the Gulf of Maine, I studied their rates of radioactive decay, which allowed me to date them back about 180 million years ago, to the opening of the North Atlantic Ocean. These findings helped support plate tectonics theory—and showed the kind of science my little undersea field jeep could do.

In 1971 Xavier Le Pichon, a marine geophysicist at CNEXO (the National Center for Exploitation of the Oceans, a French government organization), wrote to Dr. Emery that his team wanted to dive in the rift valley of the Mid-Atlantic Ridge. Part of the longest mountain range on Earth, it runs north and south down the middle of the Atlantic, from the Arctic to southern Africa, almost all underwater. Their goal was to map a 60-square-mile area about 400 miles southwest of the Azores that was thought to be a spreading center between two tectonic plates. Did we think, he asked, that submersibles could do a better job of mapping the region than sonar vehicles and other devices towed by surface ships? And could that provide further support for the plate tectonics theory?

In my mind, the French were interested in a technology transfer. Woods Hole was the juggernaut in deep submergence. We had

Alvin, and we were developing the new tracking technologies. The French wanted that knowledge. But we also wanted to learn about the geology of the ridge, and I realized that an expedition like this could be my big breakthrough.

Emery asked me to draft a reply expressing support. Jim Heirtzler, the chairman of Woods Hole's geophysics and geology department, took the lead, telling me to prepare a presentation on *Alvin*'s capabilities for a meeting at Princeton in January 1972. The giants of earth science from all over the world would be there. If all went well, the National Science Foundation would join the French in funding the project.

In oceanography, geophysicists are considered big thinkers, but they had only used sensors towed behind surface ships to map the ocean floor and determine its magnetic structure. No one had gone down to the seabed and actually seen the fresh lava flows that come out of Earth's interior, forming new crust where plates were separating. The mega-thinkers needed observational scientists like me to ground their theories in fact.

Here I was, just a graduate student, the only "Mr." in the group, invited to give a presentation on how I was using *Alvin* to gather these kinds of observations. I was down in one of Princeton's lecture pits, looking up into the heavens filled with titans, and many of them were not too friendly. It was like Daniel going into the pit with the lions. My knees buckled, but I gave my talk. When I finished, Frank Press, a geophysicist from MIT and later the science adviser to President Jimmy Carter, asked with disdain what significant piece of science submersibles had ever accomplished.

Everyone in the room stared down at me as my mind scrambled. What could I say? *Alvin* hadn't done much science yet. Then Bruce Luyendyk—a young upstart geophysicist with a Scripps doctorate who was working at Woods Hole—stood up. He was a cocky guy, and I love him to this day. He looked straight at Press and told him not to blame that on the submersibles. The scientific community had never tried to use them that deep down.

At dinner that evening, I sat across from Maurice Ewing, the founder of Columbia's Lamont-Doherty Earth Observatory. He'd had a few drinks, and he wagged his finger at me. He said he might support the project, but added, "If you fail, we'll melt that submersible down into titanium paper clips." He really threw down the gauntlet.

My competitive instincts kicked in. *We're going to do it,* I thought. The academy agreed, and the French-American Mid-Ocean Undersea Study, known as Project FAMOUS, came into being.

—

PREPARING FOR PROJECT FAMOUS required a huge shift from my dissertation research. I was jumping off the continental margin and into the deep ocean. I wasn't trained in deep-sea science, but I was excited. *Once they let us enter the Great Rift in our tiny submersibles,* I thought, *how could we miss?* We'd be the first kids on the block, the first people to drive down into this world to see the crust forming and collect pieces of Earth's new skin.

The plan was for the French to lead the show in 1973 in their bathyscaphe, *Archimède*. Given my experience with *Alvin*, I would be the only American to go with them. Then I would go out again the following year with the rest of the American team and *Alvin*, which was being outfitted with a new titanium hull that would double its diving depth to 12,000 feet—3,000 feet beyond what we needed to reach the valley.

Besides the science, the wonderful thing for me about being picked to work with the French is that I got to travel overseas. I went to France for planning meetings, and I became good friends with Jean Francheteau, one of the scientists, and Jean-Louis Michel, an engineer. Francheteau had gotten his Ph.D. at Scripps, where he had met his wife, Marta, a lovely American. He and I really hit it off. He was an earthy, handsome guy with curly hair and a great laugh—a man's man in a French envelope. He also was absolutely brilliant.

He and Marta were living in an old farmhouse in Brest, in far western France, where the French government had an oceanography center. I stayed there with them, and Jean introduced me to escargots and cheeses and pâtés—fine French cuisine, all of it. I would walk with Jean to his office, and he'd forage for mushrooms along the way. He got me ready for living aboard the French research vessel. I grew up on meat and potatoes, but when you were on a French ship, they served you tongue and stomach linings, and it was eat or starve.

And so the summer of 1973 found me at sea, hovering above the Mid-Atlantic Ridge, as Le Pichon led the first dive in the bathyscaphe *Archimède*. An older type of submersible, it dwarfed *Alvin* in size but was far less maneuverable. It just went down and up like a freight elevator. I watched the project leader complete the first day's work, knowing it would be my turn to go down in *Archimède* the next day.

I felt miserable when I woke up that August morning. I had been fighting a fever and strep throat, but I wasn't going to bail on this opportunity. I gargled warm salt water and prepared to dive.

Three of us climbed into *Archimède*. My companions were a French naval officer and a French engineer. Their broken English was infinitely better than my French, which didn't go much beyond "bonjour." The bathyscaphe floated thanks to huge tanks full of aviation fuel, which is lighter than water, and it descended by flooding its air tanks with seawater and releasing some of the gas.

Down we went. It took 90 minutes to reach the seabed, and when we got there, the view was amazing. My eyes took it all in. Lava, solidified into black glasslike shapes, ran symmetrically in both directions along the central axis of the rift valley. This landscape was made up of relatively new crust. You couldn't miss it. All of the controversy about plate tectonics evaporated just like that. The confirmation was right outside our viewport.

We maneuvered slowly and grabbed a rock sample with *Archimède*'s claw. Then something odd occurred. The vehicle's nose suddenly tilted down, and the next thing I knew we were rising at top speed. I looked at the instrument panel. The needle on the amp meter had fallen. We'd had an electrical power failure, I thought, and it must have caused the electromagnets in the ballast tanks to release some of the heavy steel shot that *Archimède* normally discharged when it was ready to float back up to the surface. As we shot up, I heard the last thing you want to hear in that situation: *"Au feu—*Fire!" At the same moment, I smelled smoke.

My companions called up to the surface, speaking in fast French. We turned off the oxygen inside the sphere and grabbed our individual oxygen masks. But even with my mask, I was getting dizzy. *Maybe I'm hyperventilating,* I thought. I started to concentrate on my breathing, hard as that was with my swollen throat. I skipped breaths to lower my oxygen levels, but my head wasn't clearing, and I was getting dizzier. *Maybe there's something wrong with my breathing apparatus,* I thought, and I began to pull it off to inspect it. Thinking I was panicking, my French companions pushed it back down, and we struggled for a few moments. Then I saw realization dawn in the pilot's eyes.

"Pardon, Bob," he said, reaching up to turn a valve that started the flow of my oxygen. That was one of his jobs in an emergency. In the midst of our panic, he hadn't done it, and I hadn't thought about it myself.

I was fine after that. But every time I saw him later, he would reach up and go, "Pardon, Bob," and pretend to turn on my oxygen. I didn't think it was funny.

All told, the French made seven dives in *Archimède* during the first part of Project FAMOUS. When I returned home, I hunkered down for a final 10-month push to finish writing my dissertation, "The Nature of Triassic Continental Rift Structures in the Gulf of Maine." My oral defense was another tense moment, with several faculty members tossing their toughest questions at me. But all the sacrifice paid off. I passed. I had my Ph.D. degree, and for the rest of my life, nobody could tell me that I couldn't do it.

—

SPIRITS HIGH, ON JULY 1, 1974, someone called *Dr.* Ballard climbed aboard *Alvin*. It was just days after I had passed my Ph.D. orals, and we were returning to the Mid-Atlantic Ridge for the second phase of Project FAMOUS. The French would be working in the northern part of the rift valley with *Archimède* and a new saucer-shaped submersible, *Cyana,* while the American team would dive farther south in *Alvin*.

Cautious not to hit any of the instruments, three of us lowered ourselves into the little sphere, just six and a half feet in diameter. Picture all of the display panels and engineering crammed in. It was like climbing into a Swiss watch. Once inside, we folded up on the floor cross-legged and slid up against the pressure hull, where each of us had a little window. With our legs commingling in the narrow space, diving became a communal experience. We joked that when we got out of the sub, we'd have to untie our legs. With my tall frame, that joke wasn't too far off the mark. We settled in with sweaters and sandwiches—diving traditions to ward off the cold and hunger during what was usually eight hours underwater.

The pilot went through the predive checks. It was over 90 degrees in there while *Alvin* was on the surface, but it would turn colder as we went deeper. When we received permission to dive, we flooded our air tanks with seawater to make them heavy and started to sink. It was a slow descent, about 100 feet a minute, or just over a mile an hour. It felt almost motionless. We were just sinking—we weren't using the sub's power to descend, because we wanted to save the battery to travel around on the bottom. Gravity was doing the work, and it took an hour and a half to drop down 9,000 feet.

People often wonder how I can contain my wild energy in such a cramped space for such a long period of time. I hyperfocus, primarily through my vision. My eyes are glued to the view outside. As we descend, that view becomes shrouded in darkness, and then I create the three-dimensional scene in my head by looking at what the instrument panels are showing me. Once on the bottom, *Alvin*'s lights illuminate the scene. I'm all eyes.

As we start to sink, all outside sounds fade, and all we can hear is the chorus of sounds of our instruments singing to us. The sonar makes a distinctive *sh, sh, sh, sh, bing, bing, bing* that changes as it detects the landscape around us. Occasionally a voice from our ship above comes in on the sound-powered telephone.

We start out in the upper layer with the sharks and dolphins. I call that the Cousteau layer—the sunlit part of the ocean that Jacques Cousteau featured in his TV shows. Cousteau spent most of his time in these shallower waters, and we like to say that we wave to him on our way to work.

Soon we enter the region where we see what we call the reverse snowstorm: particles of animals that died in the sunlit layer. Their bodies fall slowly, their remains collecting on the ocean floor at an average rate of a centimeter of thickness per millennium. These particles hardly weigh anything, so they create this snowing effect. Because we sink faster than the particles fall, the "snow" looks like it's going up. That's the only thing that gives us a sense of motion.

As we descend into the twilight zone—about 650 feet deep—the light begins to fade. By 1,000 feet, it's pitch-black. Then we see the fireworks, the bioluminescence, because the motion of *Alvin* passing by disturbs creatures that light up. We just stare out the window at these stars blinking and kaleidoscopic lights going off. It's magical, and very quiet. We drift into our own thoughts as we fall through these ocean worlds where most of the life on Earth exists. Sometimes I listen to music as I watch this world go by.

We then fall through the twilight zone, which includes the deep scattering layer, a world of strange creatures that rise at night to feed and then sink back down to avoid being eaten when the sun rises. This layer was named when a ship used an echo sounder to determine the depth to the bottom, but its sound beam bounced off the billions of tiny creatures found here.

Then we pass into the midnight zone at around 3,300 feet, where even less is going on. As we continue to fall, it's hard to imagine that we are dropping down into a mountain range. We're always thinking about climbing up into mountain ranges, not diving down into them. It's like roaming around in a balloon in the Rockies with a flashlight at night. Sometimes I would play John Denver's "Rocky Mountain High" as we dove. His voice sounded beautiful reverberating in our little sphere.

When we reach the point where there is little to see, it's time to pull out the letters we've brought along from people asking for a souvenir from the deep. Our plan is always to write responses during the dive. We write notes including our names and the latitude and longitude we're at as we write—just like what astronauts do in outer space. We'd write maybe 20 of these during a dive and mail them when we got back to land.

Atmospheric pressure is always a factor to be aware of as you dive this deep. When sitting on your porch at home or walking down the street close to sea level, you have one atmosphere of pressure on you, or 14.6 pounds of force on each square inch of your body. Once you go underwater, there's another atmosphere of pressure on you for each 33 feet that you dive. By the time we get to 3,300 feet, we have 100 atmospheres, or 1,460 pounds, pressing on every inch of *Alvin*'s hull from the outside. By the time we reach 9,000 feet, we have almost two tons of pressure on each square inch of the hull, enough to kill us instantly if anything goes wrong. But fortunately, our pressure hull protects us from calamity just inches away.

—

EVERY DAY FOR ALMOST SEVEN WEEKS that summer, we dove down in *Alvin*, staying near the bottom for five hours before we had to head back to recharge the batteries. Each time it seemed a marvel, seeing the new crust where the tectonic plates were separating. But we couldn't waste our time gawking. We took photographs, gathered rock samples, and pulled together the data to make a three-dimensional map of the area. We'd take a break for lunch—pepper steak, bologna, or peanut butter and jelly sandwiches—but once we were on the bottom, most of our time was spent at work. One major finding was that the gap where lava oozed up between the two tectonic plates was narrow, less than a kilometer wide. You could drive from one crustal plate to its opposing partner plate in a single dive.

On one of the dives, three of my colleagues were down in *Alvin* when events took a scary turn. I was on the surface ship, tracking their progress on a paper readout that plotted *Alvin*'s movements. When *Alvin* stopped to collect rock samples, the ink dots marking its location would grow into a blot. When it started moving again, the dots would turn back into a line.

After more than an hour, I noticed that the inkblot kept growing and growing and growing. *They shouldn't be staying in that spot that long,* I thought. *What's going on?* I got on a phone that transmitted sound waves through the water and called down to Jack Donnelly, the *Alvin* pilot, reminding him to move on. He said they were trying!!

Trying? My antennae went up. Walter Sullivan, a *New York Times* reporter, was on another ship, listening to our communications, and the dive team didn't want to say there might be trouble. But something was keeping them from moving.

We needed to get another submarine down to help. I radioed my counterpart on the French ship—Gérard Huet de Froberville, captain of *Archimède*—and asked how he was doing. Then I innocently inquired when he might be getting back in the water.

His antennae went up, too. He got the point. He told me he was thinking he might come over to us. I told him that would be a good idea.

As the French rescue team headed our way, the phone beeped. *Alvin's* pilot was on the line, saying they were fine and were continuing their exploration.

After they surfaced, I learned why that blot had kept getting bigger. When volcanoes erupt, they produce mounds of fresh lava. The motion of the plates pulls the lava apart and creates cracks that eventually reach the depth of a magma chamber—a pool of molten rock beneath Earth's surface. Then the chamber erupts. The crew had been exploring one of these cracks, and it looked so wide they thought there was enough room for *Alvin* to go down into it to see the internal layering of the lava flow.

But when a submersible goes down, it doesn't go straight down. It goes forward and down. Our guys went forward and down into the fissure, but they forgot that they would need clearance above them when it was time to go back up. As they had moved into the fissure, it had closed in over their head. The sub had jammed itself into the crack. They were stuck.

What saved them from a watery grave was the glassy nature of the brand-new lava above their heads. They were able to bang back and forth against it for over an hour to create more space, little by little, that allowed them finally to back out of the jam. When they came up, shards of the glassy lava were embedded in *Alvin's* side. They got some rock samples—and lived to tell the tale.

—

NEAR CALAMITIES NOTWITHSTANDING, Project FAMOUS was a huge success. The Americans made 17 dives in *Alvin*. The French made 27 dives in *Archimède* and *Cyana*. The whole project represented an important affirmation of the role submersibles could play in the deep ocean. They made it possible for us field geologists to get down and dirty, walking the ocean's terrain, so to speak.

The seafloor was spreading; we could see it. We went down to the boundary of creation, into the womb of Earth, and we documented the process taking place along the Mid-Atlantic Ridge. It was the final nail in the coffin for those who doubted plate tectonics. We could now use thousands of photos, rock samples, and our own observations to confirm the plate tectonics theory. The skeptics could no longer sneer, and *Alvin* was saved from paper clip ignominy.

My first few years of scientific research had spanned 180 million years, from the opening of the Atlantic in the Gulf of Maine to the expansion of the ocean floor today in the Mid-Atlantic Ridge. With my Ph.D. and the Project FAMOUS research, I had cut my teeth as a scientist. What a great first bite! It was certainly an unexpected start to the scientific career of an ocean geologist from Kansas.

CHAPTER 4

REWRITING THE SCIENCE BOOKS

D uring my work on Project FAMOUS, I kept up with the Sea Rovers, even taking responsibility for lining up the speakers for each year's clinic. At one of the gatherings, I told Melville Grosvenor about our research on plate tectonics and our dives in the rift valley of the Mid-Atlantic Ridge. He had retired from the National Geographic Society, but he still had sway there, and he said, "Why don't you write an article for our magazine?" He introduced me to the editors. One thing led to another, and the article, "Dive Into the Great Rift," was published in May 1975. I had entered the National Geographic family—an affiliation that has lasted a lifetime. I was delighted to have a chance to explain science to lay readers and share the excitement of exploring underwater with them.

But shortly after the article appeared, I passed a fellow scientist in a hall at Woods Hole who made a snotty dig about my publishing in a popular magazine.

"I guess you missed it," I retorted, "when we covered the same ground last year in *Nature*"—one of the world's most prestigious science journals.

I've always believed that you aren't much of a scientist if you cannot explain what you are doing to anyone who is interested, from a college graduate to a fifth grader. But that was not the view at most academic institutions, where scientific purists care only about publishing in peer-reviewed journals that add the most luster to their reputations. I had learned to play that game in getting my Ph.D., and now, in 1975, I was under greater pressure to make big discoveries—and publish them in science journals—to gain tenure and, with it, freedom to do whatever I wanted in science.

I was 33, and I had to earn tenure by the time I was 38 to continue as a scientist at Woods Hole. Its program was the gold standard, and the bar was very high. The majority of the postdocs who came there did not end up staying. I had spent three years in the Navy, and although I got to do some research then, I was behind my peers. Even at Woods Hole, with all its Navy work, the attitude was, "Tough. That was your decision." I never expected to be rewarded for my military service, but I certainly didn't expect to be punished for it. I was really going to have to double up. I remember thinking, *My God, am I going to make it?*

I started 1976 with a dive in the Cayman Trough down to *Alvin*'s new limit of 12,000 feet, exploring a spreading center between two plates in the Caribbean Sea that contained the deepest volcanoes known at that time. I had just received my first National Science Foundation grant as a principal investigator to study these plates, which were grinding past one another like the San Andreas Fault, letting molten lava pour into the void. This was the first phase of a multiyear study, and I didn't see why I couldn't be both a scientist writing for refereed journals and an explorer sharing the magic of discovery with the general public. So I brought Emory Kristof, a National Geographic photographer, and Walter Sullivan, a *New York Times* science writer, along with me. I dove into the trough with other scientists, taking pictures and retrieving rocks that could tell us more about the structure of Earth's crust.

—

PARTWAY THROUGH THAT PROJECT, though, I took a short break for a different underwater quest that probably drove my more traditional colleagues crazy: I helped *National Geographic* on an article about the Loch Ness monster.

Emory Kristof said I should apply my scientific know-how to the hunt for this mythical beast, and I thought, *Why not?* There were good reasons to take up the challenge. I had never seen the Scottish Highlands, and this was a chance for Margie and me to do it on someone else's nickel. We also needed the money that the *Geographic* project would provide. And, honestly, sometimes it helps not to take yourself too seriously—or so I told myself. So Emory and I joined a team of divers, using sonar and underwater photography to make images of the lake's bottom. We also dangled bait from a specially designed rig to lure the monster into view.

I reasoned that if there was a Loch Ness monster (singular), there had to be Loch Ness monsters (plural). The first reported sighting of Nessie dated back to A.D. 565. Then a mysterious photograph taken in the 1930s purported to show the monster. So if this species had survived some 1,300 years in Loch Ness, there must have been a minimum of 25 Nessies over that time to sustain the population. That meant there should be lots of Nessie bones, because bones don't dissolve in freshwater the way they do in the deep sea. I dredged the floor of the loch but turned up only tree trunks and things like a shoe and a teakettle.

I did, however, discover an interesting phenomenon: If I navigated straight down the long axis of the loch, the ship's wake would roll out toward the rocky walls on either side and bounce off them. Long after we had passed, the two returning waves would meet back in the middle, creating a standing wave consisting of two to three humps that would last for a few seconds and then disappear. Could that have been what people had been seeing?

As I bid farewell to the Highlands, I stopped at a roadside lunch stand to grab one more "Monster Burger." I had spoken with this vendor at the start of our adventure, and he had begged me not to sink the local economy by disproving the Nessie legend. Now I was able to reassure him. "It is impossible to prove an animal doesn't exist," I told him, "if it never existed in the first place." It was a fun trip, ending with a visit by a film crew for the television show *In Search Of,* hosted by *Star Trek*'s Leonard Nimoy. I'm sure that also went over well with my scientific colleagues.

—

IF I WANTED TO GET TENURE, though, I needed to prove things, not disprove them. With the success of Project FAMOUS and our first trip to the Cayman Trough, more scientists were recognizing the value of our highly visual approach to ocean exploration. Some were asking me to join their expeditions, as long as I brought along the tools of my trade, not only *Alvin* but now also *Angus* (an acronym for Acoustically Navigated Geological Underwater Survey)—a camera sled that we could dangle down to 20,000 feet on a long trawling wire. *Angus* was originally a simple "dope on a rope"—a black-and-white camera enclosed in steel rails that protected it in volcanic terrain. I had added two larger color cameras and tested them on the fresh lava flows in the Cayman Trough, and now *Angus* could snap thousands of photos that we developed as soon as we hoisted it back up onto the mother ship. I also needed to form a team, headed up by Cathy Offinger and Earl Young, who were at sea with me for all of the major discoveries I would make while at Woods Hole.

Word spread about my ability to get up close and personal with the sea bottom. In early 1977, I was invited to join a team from MIT, Stanford, Woods Hole, and Oregon State to dive about 200 miles north of Ecuador's Galápagos Islands, where Charles Darwin had come up with seminal theories on evolution. Jack Corliss, the leader of the expedition, had dived with me in the Cayman Trough, and he really wanted to use *Alvin* to explore the Galápagos Rift, another tear in Earth's crust. He and other scientists had collected sediment and rock samples on the flanks of the rift that could only have been generated by hot water circulating through the upper crust of the ocean floor on its way to the surface. That suggested these plates were separating more rapidly than the ones we had studied along the Mid-Atlantic Ridge and that more heat was escaping. This raised the question: Could there be vents in the seabed where warm springs—or even much hotter ones—were rising up from beneath the ocean floor?

A year before, a team from Scripps had scanned the area, which was 8,000 to 9,000 feet deep, using *Deep Tow,* a sonar vehicle, and had detected a few subtle temperature anomalies. *Deep Tow* operated much higher off the ocean floor than *Angus,* so it seldom got close enough for good images. I knew we could get down in *Alvin* for a much closer look, but the real ace up my sleeve would be *Angus*—the photographic scout that could cover much more area than *Alvin* and tell *Alvin* where to go.

As we prepared for the Galápagos dives, we added a temperature gauge to *Angus* so if it picked up an unusual reading, we'd know right away where we should concentrate our efforts. We lowered *Angus* to within 13 feet of the seabed and towed it slowly, at less than two knots, for miles and miles, taking thousands of pictures over a 12-hour run. At one point, we saw a blip on the temperature scroll lasting three minutes. Clearly, we'd just gone through warmer water.

Once we got *Angus* back on the ship and processed the film, we focused on the photos taken during those three minutes. So we were sitting there, going through photos of fresh lava and then more lava and piles of broken rocks. Then, all of a sudden, as we're getting close to that three-minute blip, the water in the frames gets cloudy. Then we see all these clams, big white ones, sitting on solid rock. *Clams?* There weren't supposed to be any clams down there, not in a world of total darkness with little food to eat and resting on solid rocks. What was going on?

I had seen occasional fish in the deep that lived off tidbits of organic matter sifting down from the sunlit layers. But this seemed like something entirely different—a whole ecosystem thriving where no one knew one could exist.

Now we knew where *Alvin* should go.

Jack Corliss and Tjeerd Hendrik van Andel—a friend of mine from Stanford whom we called Jerry and who also had been part of Project FAMOUS—went down in *Alvin* on the first dive. As the clams came into view, the temperature readings jumped from a typical 2.5°C (36.5°F) to 16°C (61°F). The water turned light blue in *Alvin*'s floodlights.

"We're sampling a hydrothermal vent," Jack called out through the underwater phone. He and others had used this term in their theoretical writings, but this was the first time anyone had ever seen one, a crack through which hot, mineral-rich water was spewing out from beneath the ocean floor.

Baffled that a colony of clams could be living here, Jack added, "Isn't the deep ocean supposed to be like a desert?"

And there was wilder stuff to come. We dove the Galápagos Rift 20 more times over the next five weeks and found other areas that were teeming with life, like oases in the deep.

One was populated by large brown mussels, another by flower-like creatures, and a third by clusters of giant tube worms that looked like rose hedges. None of us was a biologist, so we named them after whatever came to mind: spaghetti worms, the dandelion patch, the garden of Eden. They were stunning and exotic. We were finding whole communities with species and food chains that weren't even in biology books. But why were they here? No life as complex as this should exist, we thought, without the light and warmth of the sun. None of it fit anything we'd been taught.

People talk about eureka moments, when scientists know they have discovered something big. In our case, the first clue about how this strange ecosystem worked came one night at dinnertime when John Edmond, a chemist from MIT, opened water samples brought up from a vent. An odor like rotten eggs spread through the ship. We were coughing and running around opening windows. Oh my God, it was powerful. We knew that smell: hydrogen sulfide. That stuff will kill you if you inhale too much of it.

When we opened some of the clams that *Alvin* had plucked from the bottom with its mechanical claw, we found a mass of bloodred flesh rather than the gray inner bodies we were used to in our Cape Cod steamers. The flesh looked like raw steak, and was bleeding like when you cut yourself. The clams had no internal organs. Just this mushy stuff.

We needed to bring samples back for biologists to examine, but we hardly had any formaldehyde to preserve them. So we improvised. The ship was supposed to be dry—except for beer—but some members of our team had their own secret stashes hidden away in their cabins. Out came the gin and vodka, and we used that to preserve our discoveries.

When scientists ashore looked at the clams under microscopes, they discovered that they contained a large mass of very primitive bacteria, possibly a type that had formed when life on Earth first began to evolve into more complex forms. In very simple terms, you might say that early in the evolutionary process, these bacteria had struck a deal with the clams, saying, in effect, "Do me a favor and let me live in your body. That way you can inhale this poisonous hydrogen sulfide, which should kill you, but it won't if you share it with me."

It was quite a bargain. We had all been taught that in the absence of sunlight, you cannot have photosynthesis, and without photosynthesis, you couldn't have complex ecosystems. Wrong. Up to that point, no one knew about these crazy bacteria and their role in another chemical process that we now recognize as a basis of life: chemosynthesis. Here we were seeing one of the cradles of life on our planet that no one had imagined before.

A couple of years later I went back to the Galápagos Rift with a group of biologists, who confirmed that the bacteria living inside the clams and tube worms were able to perform a process analogous to photosynthesis in the dark. They use molecules in the vent fluids as a source of energy and convert them into substances that sustain the animals. The red mushy interior of the clams and the red tips of the tube worms were filled with hemoglobin, the building block of human blood. Oxygen was in low supply in the vent fluids, and the hemoglobin stored it, helping to fuel the gigantic clams and other creatures. The biologists were just as astonished as we had been that the bacteria could support a complex community in such a harsh environment.

In the years since, scientists have come to recognize that these communities hold critical clues as to how life evolved on Earth— and could evolve on other planets. It's the discovery that is now driving NASA's efforts to explore moons revolving around Jupiter and Saturn. They have ocean worlds beneath their ice surfaces that are far larger than our own oceans, and perhaps they have life-forms unknown to us.

—

JUST MONTHS LATER, in June 1977, my team returned to the Cayman Trough—the colossal opening in the Caribbean Sea floor that stretches from Guatemala to Jamaica—to gather lava samples from the active volcanoes there. Those were the deepest dives I've ever taken—to depths greater than 20,000 feet. We were diving in *Trieste II*, a Navy bathyscaphe similar to *Archimède*, the French craft we used during Project FAMOUS.

This second mission to the Cayman Trough got off to a bad start. I flew to Panama, where I was to board a Navy vessel that would take us out to sea. It wasn't until I was walking up the gangway that I realized I had left a set of sensitive Navy maps in the overhead bin on the plane. I flagged down my cabbie, who was in mid-three-point turn, and we sped 21 miles back to the airport. It was closed for the night. This was back before 9/11, when airport security wasn't that tight. Still, soldiers were walking around with automatic weapons. My only thought was those maps, which were critical for the expedition. I walked right past the soldiers, cussing myself in English as a cover, and back onto the aircraft, right past the cleaning crew on board. There were the maps, untouched in the overhead bin. Another 21 miles in the cab, back up the gangway, and I was ready to go to work, maps in hand.

Trieste II had a proud history and had gone through many modifications and upgrades. It had surveyed and photographed the wreckage from a nuclear submarine, U.S.S. *Scorpion*, that sank in 1968, and it had recovered sunken satellite gear in 1972. At 70 feet and 88 tons, *Trieste II* was a long, heavy tube. Like the French bathyscaphe *Archimède*, its buoyancy relied on tanks of aviation fuel. In 1977, it represented outdated technology, largely overtaken by nimbler craft like *Alvin*. But it could take us down to 20,000 feet—8,000 feet deeper than *Alvin* could go.

We were on the last dive, and it had been peaceful enough that I'd been able to catch a long nap during the nearly six-hour descent. I was looking out of our only viewport as the pilot called out that we were 200 feet from the bottom. All of a sudden, I saw the side of a volcanic wall angling down below us. Somehow our bottom-finding sensor had failed to pick it up.

"Bottom!" I yelled. "I see bottom."

The pilot quickly began releasing some of the iron pellets that formed our ballast, trying to slow our rate of descent. But bathyscaphes don't stop just like that. And with a sickening grinding noise, our bow gouged the volcanic scarp. The rocky wall retaliated by bending a steel girder in our bow.

Now I could visualize what had happened. As the bow of our long vessel had crossed over the bottom slope of the wall, the stern—where our bottom-finding sensor was mounted—still jutted out over the deeper depth.

But that wasn't the worst of it. As I looked out the tiny viewport, I could see something colorful in the water. It looked like a mirage. It's the effect you get when you mix gasoline or oil with water—and that's exactly what I was seeing. It was aviation gas. That could only mean that we'd ruptured one of our flotation tanks, and we were bleeding out the gas that was supposed to get us home.

The pilot immediately released all of our ballast, aborting the dive. We started to rise. But it would take six hours to reach the surface, and now the question was: How fast was the gas leaking out? Were we going to bleed out all the gas before we got to the surface and end up making a one-way trip back to the bottom? We didn't know.

We stared at the LED readout showing our rate of ascent. But the numbers bounced around from one calculation to the next. Were we slowing down, or were we not slowing down? It depended on whether you were an optimist or a pessimist.

For six hours, no one said anything. We knew we had to get to the surface on our own. We were so deep, there was no question of anyone coming down for us. Finally, as we could see the sunlight penetrating the water, we heard a distant *clunk-clunk*—the sound of the rubber Zodiac coming to meet us as we broke through the surface. I still remember the feeling of that first breath of fresh air I gulped when we finally poked up through the hatch.

That's when I started to think, *Why am I doing this?*

I wouldn't say I lost my nerve. Despite the dangers, I did dive many more times in submersibles. But I did start thinking, *Why play Russian roulette? There's got to be a better way.*

—

THE FACT IS, I was always thinking about whether there was a better, and safer, way to explore. I remember once diving in *Alvin,* when we had located a vent with giant tube worms. I glanced over at one of the biologists, Holger Jannasch, and he wasn't even looking through the viewport. He was gazing over my shoulder, watching a four-inch TV monitor that showed what *Alvin*'s camera was filming. Really? Do any of us need to be down here or even on the ship above? Couldn't we build robots to do all this remotely?

That idea was starting to form in my head, but for the time being it seemed clear that my best bet for tenure was to learn even more about hydrothermal vents, maybe how to predict where we might find them. It seemed likely that they could be found along the mid-oceanic ridges. I decided to focus on creating a model to predict where they might occur.

In 1979, my team joined an expedition led by Scripps about 100 miles south of Mexico's Baja California Peninsula on the East Pacific Rise, the ridge that crosses the Pacific Ocean floor from Mexico to where it enters the Indian Ocean south of Australia. I had worked along this rise a year before with French scientists from Project FAMOUS, led by my friend Jean Francheteau, who was joining the expedition as well. It made me nervous that the *Melville* cruise was led by my own personal Ahab, Fred Spiess, who had rejected me for graduate school 14 years earlier. I doubted he remembered me, though, and I did not remind him of that history.

Spiess and most of the other scientists on board were researching volcanic and seismic activity, while Jean and I were focused on finding vents. But there was only one winch on board for lowering vehicles into the sea, and Spiess insisted that his *Deep Tow* sonar vehicle go first. *Angus* had a strong, protective steel cage around it, but *Deep Tow* was more vulnerable, like a precision race car. And when Spiess's team banged into some undersea cliffs, it had to be hauled in for repairs.

That gave me a chance to slide *Angus* into the water, and we quickly started finding evidence of hydrothermal vents. This time, though, they looked different from the ones we had found in the Galápagos Rift. The water coming out of the vents was cloudier, almost like black smoke, so we sent *Alvin* down to take a look.

Dudley Foster was piloting *Alvin* as it maneuvered into a thicket of strange chimneylike structures, approaching one that was spewing black clouds of suspended minerals.

"We've got a locomotive blasting out this stuff," Dudley called up to us. It looked like the smoke from a coal-burning engine. As he neared that chimney, *Alvin*'s probe was recording a sizzling 33°C (91°F)—much hotter than the temperatures we had found around the vents in the Galápagos. It was so hot, the tip of *Alvin*'s thermometer melted.

That evening, Jean and I talked about what these smoke-spewing chimneys could be. Two years earlier, in the Galápagos, we had come upon similar tubular deposits, but they had crumbled when I tried to grab a piece with *Alvin*'s claw. And in 1978, on a cruise Jean and I had conducted with the submersible *Cyana,* a

Mexican scientist had seen a chimneylike structure and had taken a rock sample. The sample had sat in a French lab for months until an American scientist identified it as a sulfide rich in zinc, copper, lead, silver, and gold—a mineral substance that had never been found on the seabed before. But nobody had ever come upon a smokestack scene like the one we were seeing. It seemed to come right out of the industrial revolution.

The next day, Jean and I made sure a gauge that could record much higher temperatures was installed on *Alvin,* and we dove down to the same location. We had to see it for ourselves. Some of the chimneys were six feet tall, and they were spewing out vast black clouds of smoke. "They seem connected to hell itself," Jean said.

As we drew near one that was belching, our pilot suggested we take a temperature reading. Thank goodness he did. We were flabbergasted to see the gauge shoot up to 350°C (662°F), three times the boiling point of water. We quickly backed off. Our titanium hull could have withstood it, but any closer and the heat would have cracked our Plexiglas porthole and created a massive and deadly implosion. We had calculated that these high-temperature fluids could be generated by the magma chamber some kilometers beneath the ocean floor, but we never expected these vent fluids to maintain that high a temperature as they came out into the ocean water. But clearly they did.

On subsequent runs, *Angus* found more smokers, and we vectored *Alvin* in to inspect them. We had just installed an automated acoustic navigation system that could be operated by a single individual, compared to the three- to four-person navigation team that *Deep Tow* needed. We were feeling pretty good about where we stood at the moment in the long-standing rivalry between Scripps and Woods Hole, and clearly my competitive nature came out, so I told Steve Gegg, who was navigating *Angus,* "When you're on watch, I want you to play some music, pop some popcorn, and put your feet up on the computer." That kind of casual attitude would drive Spiess—a by-the-book former Navy officer—insane. And each time *Angus* found another smoker, I had my team stencil a small picture of one on *Angus*'s side, sort of the way they'd paint a rising sun on the side of World War II fighter planes for every Japanese Zero they shot down.

I had invited my father to come on that cruise with us. One evening he was on an upper deck and heard Dr. Spiess come out into the night on the deck just below him and let out this big exhale, almost a scream. This was a buttoned-up guy who rarely lost his temper, and I'd clearly gotten under his skin.

In subsequent dives, our team found more chimneys, some almost 30 feet high. We called them "black smokers" or "white smokers," depending on the color of their emissions. All of them were made up of polymetallic sulfides, the same minerals in that sample the French had collected the year before. When we looked carefully at the fluid coming out of the chimney openings, we could see it was clear. But the moment it began mixing with the cold bottom water—4°C (39°F)—the fluid quickly quenched, or cooled, and turned black as the minerals it contained formed micro-crystals. We began to understand how these chimneylike structures formed. As hot fluid spews out of the newly fractured rock, it mixes with the ocean water. As the minerals precipitate around the edge of the opening, the chimney gets taller and taller until it gets top-heavy and falls over, only to begin the process again. The result is a large mound of minerals combining with many separate chimneys and looking almost like a pipe organ.

Our discovery helped explain the chemistry of seawater. In chemistry classes, we learn about the hydrologic cycle. The warm sun causes water to evaporate from the ocean. This pure water forms moist clouds, from which rain falls onto land and flows through rivers back into the sea, picking up chemicals as it goes, only to be evaporated over and over again.

But the chemicals in river waters are different from what we find in a typical bucket of seawater. Some of the chemicals flowing into the sea strangely disappear, while new ones take their place in the ocean water. But where do they come from?

Once we had discovered these high-temperature black smokers, we realized that a second circulation system was involving the entire volume of the world's ocean and circulating inside Earth's ocean floor over millions of years. Within the sulfide deposits making up the chimneys, we also found copper, lead, silver, zinc, and gold—valuable metals in large enough quantity that it helped set off an underwater gold rush just now playing out.

—

PLATE TECTONICS, hydrothermal vents, and black smokers. In very short order, we—all the scientists that I sailed with, French and American—had rewritten parts of the geology, biology, and chemistry textbooks we had studied in college. Project FAMOUS threw away our geology book; discovering the vents threw away our biology book; and finding the smokers threw away our chemistry book. It was an amazing trifecta. Many of the scientists knew more than I did about the theories behind the discoveries and the impact they would have. But I was fortunate enough to have had the right tools and skill set to get them to the right places. Without *Angus*'s cameras and *Alvin*'s viewports for scientists to peer through, we wouldn't have seen the proof with our own eyes.

I soon accepted an offer from Dr. van Andel to spend the 1979–1980 school year on sabbatical at Stanford, where I could work on academic articles about our latest discoveries. I was up for tenure, and I certainly felt like I deserved it now. Tenure at

Woods Hole was determined by a panel of outside experts, and when I heard who was on the committee that was going to decide my fate, a knot began to form in my stomach: some of my biggest rivals, including—you guessed it—Dr. Spiess. I had just gone toe to toe with him at sea, and now I felt like I had a crosshair right on my forehead.

I decided that Margie and the boys should move with me to California. We would rent our Hatchville house. That way if I didn't get tenure, we could just stay on the West Coast, in the state I had grown up in. I had been spending a lot of time away from home, but I remembered all the fishing and camping trips that my dad had taken us on, and I made it a point to take Todd, who was now 11, and Dougie, who was turning 9, on weekend trips like that. I also coached Todd's basketball team, and we got to spend some time swimming and fishing at our favorite vacation spot, a cabin that Margie's father owned on Whitefish Lake in Montana.

Still, I was feeling more turmoil about my marriage. As at Woods Hole, Margie had no interest at Stanford in socializing with my professional friends or sharing in my intellectual life. I watched my brother, who now was working in software at Apple, as he and his wife were going through a divorce. I could see how hard it was on their three children, and I decided that I would try to make things work with Margie as long as our boys were still living at home.

I was also still feeling some disdain from other academics because I had continued to write for *National Geographic*. I had published magazine articles on the Cayman Trough, the oases around the vents, and the smokers. I was working on my first National Geographic Television special—about plate tectonics and our discovery of the vents and smokers—and I could already hear the cries that I was sure would come from some of my colleagues, saying I was just showboating. But I also had co-authored major articles in two top academic journals, *Nature* and *Science,* and had two more on the way, one of which would end up winning the Newcomb Cleveland Prize for the best article in *Science* in 1980. Critics scoffed that I liked to go out and just see what I could find. They said it wasn't science—that I just tripped over things. *If that's the case,* I thought, *then let me keep tripping.*

Luckily, with all this weighing on me, the chairman of my tenure committee was Robert W. Morse, who had been the Navy's assistant secretary for research and development and now held a high-level post at Woods Hole. He also was my tennis buddy. Later I learned from him that the committee had sought letters about me from other experts, including some who had never said a nice thing about anyone. They included scientists who had just been in the field with me, like Jerry van Andel at Stanford, Harmon Craig at Scripps, and John Edmond at MIT.

It came down to a vote, with everyone sitting around a table on February 14, 1980. I was not allowed to be there, but Morse told me later that when it got to Spiess, he went off on what a publicity hound I was.

Then Morse asked, "If Ballard were at Scripps, would you give him tenure?"

And Spiess replied, "Yep."

What a relief! I had finally made it. I was now a tenured scientist at Woods Hole and free to chart a new destiny for myself and my team.

CHAPTER 5

MAKING FRIENDS WITH THE NAVY

W hen I'm working on one thing, I'm always dreaming up
the next, just waiting for the right opportunity. I'm like
Huck Finn sitting on the riverbank with strings connected
to the bobbers on my fishing lines. I never know when I'll
get a nibble that will lead to the big one. So while I was absorbed
in my academic research on the way to getting tenure, I was also
thinking about better ways to explore, using the latest in robotic
technology. And I still held on to what my Sea Rovers friend had
called, more than 10 years before, my pipe dream: finding *Titanic*.

Titanic had been on her maiden voyage, traveling in the
spring of 1912 from Southampton, England, to New York City.
She was the grandest ship of her time, almost three football fields
long, and just as spectacular by any other measure. Besides all
the chandeliers and fine wines and dining, first-class passengers
could enjoy a deluxe suite, as well as a gymnasium, a squash
court, a Turkish bath, and smoking and writing rooms, all for
$4,350—about $115,000 today. Some of America's wealthiest
were on board.

But exuberance bred overconfidence. As the ship steamed through the North Atlantic, several hundred miles southeast of Newfoundland, the captain, Edward J. Smith, didn't seem alarmed by warnings about icebergs in the area. Late at night on April 14, a lookout high in the crow's nest spotted a gigantic iceberg dead ahead. The starboard side of the hull scraped against the ice below the waterline as the crew frantically maneuvered to avoid the hit. Compartments below flooded, dooming the ship, which finally plunged beneath the surface at 2:20 a.m. on April 15, just two hours and 40 minutes after the calamity had begun.

The tragedy became the biggest news story of its time, marking the end of a prosperous era before World War I and capturing the public's imagination ever since. For ocean explorers, finding *Titanic* was like climbing Mount Everest, the number one quest on everyone's list, and that's why it held such allure for me. No one had ever searched for the wreck, which lay in water up to 13,000 feet deep. A company called Big Events had contacted me in early 1977 about looking for it. Through them, I met William H. Tantum, the president of the Titanic Historical Society, an organization devoted to learning about the ship and its passengers. Bill was a sweet guy and a vivid storyteller, a Yankee version of Shelby Foote, the southern historian in Ken Burns's epic Civil War documentary. When Bill was talking, it was like you were on *Titanic* with him. He let me look through all the books, maps, and drawings he'd collected, and his passion to find *Titanic* stirred my own. We backed away from Big Events when we learned that the company wanted to market paperweights from pieces of *Titanic*'s cables, but Bill and I stuck together and looked for other opportunities.

Some officials at Woods Hole cautioned that pursuing a project with such popular appeal was beneath the dignity of a serious research institution. But Paul Fye, who was nearing the end of his tenure as Woods Hole's director, took the view that *Titanic*'s resting place in the northwestern Atlantic made us the logical institution to lead the search for its wreckage. He didn't want Scripps coming over from California and finding it in our backyard, and I certainly agreed with that.

That same year I started talking to Alcoa about using a ship, *Seaprobe,* that had been built for another explorer, Willard Bascom, to search the Black Sea for ancient shipwrecks. Bascom thought a deep layer of water devoid of oxygen might have preserved wooden ships there, and he wanted to try to raise some of them. Given Cold War tensions, though, he'd never gotten to the Black Sea, and *Seaprobe* stood idle. Alcoa wanted to donate the ship to Woods Hole.

You might say that *Seaprobe* was the perfect ship for pursuing a pipe dream. It was essentially an oil drilling derrick mounted on a sturdy aluminum hull, and it had the machinery to lower lengths of drilling pipe holding either cameras or a giant claw to the seabed. In fact, it was similar to *Glomar Explorer,* a larger vessel that the CIA had recently used to retrieve part of a sunken Soviet missile submarine in the Pacific. I just wanted to find, not raise, *Titanic,* and attaching a pod with cameras and sonar to the bottom of *Seaprobe*'s pipes would give me a chance to test some of the latest devices.

"Let's tell Alcoa to bring the ship up to Woods Hole, and you take it out and run it through its exercise," Dr. Fye said. "See if it's worth it."

With his green light, I started borrowing the gear I'd need, including a sonar system from Westinghouse and underwater cameras from the Navy and a local company called Benthos. All in all, it meant we were going to be dangling a pod containing $600,000 of equipment on the end of a very long pipe.

We sailed *Seaprobe* into the Atlantic in October 1977. The riggers did well the first day, easing one 60-foot length of pipe after another into the sea. By nightfall, they had attached 50 lengths, lowering our sensor pod to 3,000 feet—just one pipe length above the ocean floor.

The riggers knocked off for the night, but I stayed up, working in the control center to begin our first sonar run. Then, all of a sudden, at 2 a.m. we heard an ear-splitting crash above us, and the vessel trembled. Everyone rushed out to see what had happened. A connection between two pipe lengths had broken somewhere. A giant counterweight atop the drilling derrick had slammed down onto the upper deck above us while 30 tons of pipe smashed into the seabed below, obliterating the pod with all the equipment I had just borrowed.

We quickly realized why the accident had happened. Just before we'd sailed, our drilling contractor had quit in a pay dispute, and his inexperienced replacements didn't realize that the last section of pipe they added needed to be superthick to keep the whole assembly from bending and breaking as the ship moved through the water.

For want of a few extra bucks on the front end, we'd lost the $600,000 of equipment I'd borrowed, along with any chance to use *Seaprobe* to find *Titanic*. My pipe dream had turned into a pipe nightmare. What made it even worse was that Dr. Fye—my mentor, friend, and protector—had just retired as Woods Hole's director. An interim team was in charge and in no mood to make good on my costly failure.

"You're in trouble, Ballard," one of them said when I returned. "We didn't authorize you to borrow all that stuff. You did that on your own, and now you're going to have to pay for it."

I had about four dollars in my pocket, and my old, partially renovated farmhouse was not worth a lot more than the $20,000 I'd paid for it. It looked like I was facing financial ruin. Then Alcoa, thank God, came to my rescue. The company's insurance policy covered not only the ship but also the equipment I'd borrowed.

Still, such a highly visible failure was crushing to me. When the new director, John Steele, arrived at Woods Hole, he said he did not want me to do anything else under the institute's name to look for *Titanic*. A Scottish oceanographer, Steele was an academic purist, and he insisted that I stick to basic scientific research.

—

I'VE NEVER BEEN AFRAID OF taking chances, and I wasn't going to stop now. My entrepreneurial instincts kicked into gear, and I figured out how I could continue my quest for *Titanic* on my own time. A group of us got together—Bill Tantum; Emory Kristof, the National Geographic photographer; Alan Ravenscroft, a filmmaker—and formed a company to try to find *Titanic*. We talked to wealthy individuals, including Roy E. Disney, Jr., Walt Disney's nephew, but everyone seemed to think the idea was too risky. A Texas oilman, Jack Grimm, suggested we work with him, but he was hungry for publicity and didn't strike us as a reliable partner.

So I decided to spend part of my sabbatical at Stanford refining my ideas on how to conduct deep-sea searches. The personal computer revolution was under way, and others at Stanford were working closely with people like Bill Gates and Steve Jobs, testing robots in the quad right outside my office. I was imagining a similar paradigm shift in my own field. I was moving away from manned submersibles, which were dangerous and could stay underwater only a few hours at a time, to underwater vehicles that could be operated from on board a mother vessel and that could remain submerged for as long as needed. I even gave names to the robots I was envisioning. I planned to call them *Jason* and *Argo*, in honor of the mythical explorer and the vessel in which he had brought home the Golden Fleece. Compared to *Alvin*, they would be cheaper to operate and could survey much larger areas—a critical factor given the strict time limits on most ocean expeditions. I also envisioned a day when these systems could stream photos and videos anywhere in the world via satellite, enabling scientists to participate in discoveries from their offices through telepresence.

My experience with *Alvin* had shown me that it takes 10 to 15 years to prove out a technology before it gains wide acceptance. Scientists were now lining up to dive in *Alvin,* telling their friends that they were risking their lives for science. I knew it was time to move on.

But I still had to convince others that my ideas on remotely operated vehicles made sense and were worth funding. The National Science Foundation said no. I love how people like that say, "This idea you've got is crazy, though your last one was pretty good." And I say, "But don't you remember that you called that one crazy back then, too?"

Finally I turned to the Office of Naval Research, which was Woods Hole's biggest sponsor and had supported me since the start of my science career. My current grant application was sitting there when Dr. Morse, the head of my tenure committee, asked if I could speak at a Navy event, the International Seapower Symposium at the Naval War College in Newport, Rhode Island. I was still in the Navy Reserve, and I enjoy public speaking, so I said sure. As the date approached—June 1981—I asked my secretary, Terry Nielson, for the details. She said I was expected to stay for four days. Four days talking to a bunch of engineers about shipboard power plants? No, she said. That's not it. This is a gathering of the top admirals from all of the navies in the free world, and they'll be talking about projecting military power at sea.

President Reagan had just taken office, and he and his brash young Navy secretary, John Lehman, were making plans to confront the Soviets more aggressively. Reagan had promised to expand our Navy from 450 to 600 ships, and he and Lehman wanted to rattle the Soviets psychologically, making them think we had so many capabilities that they could never match us in a war. Lehman was going to be one of the speakers at the symposium. So was Adm. Thomas Hayward, the chief of naval operations. And guess who would be the other main speaker? Yep, me.

Oh, my God, I thought. What have I done? I'm a marine geologist, not a defense strategist. I feared I would make an idiot of myself. After I willed myself to calm down, I thought, *Go with your unique strengths and just put a Navy spin on them.*

I knew from my Army training that terrain was everything; you always wanted to take the high ground and embed yourself within it. What was the high ground in the ocean? Clearly, it was the mid-ocean ridge, which runs around the planet like the seam of a baseball, including the terrain under the Atlantic Ocean that I had explored in Project FAMOUS. Off my mind went, dreaming. I blended the two concepts together, and after a trip to Woods Hole's graphics department, I was ready.

With Lehman and Hayward sitting in the front row, I spun out my concept of undersea terrain warfare. I showed how Soviet submarines heading toward the eastern United States had to cross over this mighty underwater mountain range. One slide showed submerged peaks and outcroppings that I had labeled "SAM sites," an Air Force and Army term for land-based surface-to-air missiles that I used to get their competitive juices flowing. These sea-based sites could be filled with mines that could fire torpedoes at Soviet subs, I suggested. I showed them an ecosystem of large and small submarines that could hide in the rock fissures and launch them remotely. Everyone in the room loved it when I said the Bad Guy would never know what hit him.

I met Lehman at a reception that evening, and we hit it off right away. He grew up in a wealthy Philadelphia family—first cousin once removed to Grace Kelly, the movie star who became princess of Monaco—but he was a scrappy, down-to-earth guy who had served in the Navy Reserves as a navigator on attack planes. We were both just turning 39, and we both liked to disrupt things. He was already giving instructions to the admirals in ways that few Navy secretaries had done.

I invited Lehman to dive with me in *Alvin,* and he said he'd love to. As I headed home, I was thinking about how good it felt to reconnect with the Navy. I also was wondering if Lehman and Hayward could help me get funding to develop my remotely operated vehicles and maybe even get another shot at finding *Titanic.* But I knew that a lot of things would have to fall into place for that to happen.

—

ON THE *TITANIC* FRONT, the first hurdle was the one that Dr. Fye and I had feared the most: competition. Jack Grimm, the self-promoting oilman, had managed to recruit Dr. Spiess from Scripps and William Ryan, a marine geologist from Columbia University, to help him. Spiess and Ryan, a formidable duo, first went out looking for the ship in 1980, but they were hindered by equipment problems and terrible weather, not to mention Grimm himself. He had hired someone to train a monkey to point to a spot on a map where he thought *Titanic* was, and he wanted to send the monkey out on the ship. Spiess and Ryan had said, "It's either us or the monkey," and got him to drop the stunt.

But now Grimm, Spiess, and Ryan were mounting a second try, and I was waiting nervously for the results. I learned later that they spent a lot of time studying fuzzy sonar targets from the year before and following a hunch that the ship might have slid into a vast underwater canyon. Then, in the final hours of the expedition, Grimm seized on a shadowy image that looked like a propeller and announced publicly that he'd found *Titanic*. My heart almost burst when I heard the news. But it turned out that neither Spiess nor Ryan supported Grimm's theory, and his story fell apart.

I didn't think Grimm had enough credibility left to make another attempt. But I was feeling more urgency to find funding for my undersea robots, and I had a feeling that it was going to take all my sales skills and persistence—and a deeper entrée into the highly classified world of undersea warfare—to persuade Navy officials to help me with my *Titanic* dream.

Later that summer, Admiral Hayward called and invited me to the Pentagon to give my terrain warfare presentation to his staff. Vice Adm. Ronald Thunman, the top submarine officer, also wanted to see me. That seemed like a good start. Off I went down the E-Ring of the Pentagon, the corridor with the nicest offices for top officials, carpeting and windows and all. Thunman's office, however, was more like a prison cell—in a secure area, with bars on the office door window. As the door closed behind me, Thunman did not even look up.

Then he exploded out of his chair and began screaming, "What the hell do you know about submarine warfare?" Thunman was a towering man, and it seemed like flames were coming out of his mouth. Leaning back from all this shock and awe, I thought to myself, *I don't need to take this. I didn't ask to be here.* So I counterattacked.

"What the hell do you know about the bottom of the ocean? That map on your wall tells you nothing," I said. "Just that the ocean is blue. Don't you realize it contains the largest mountain range on Earth, of critical importance to the future of the Navy?" I asked if he realized that either he—or the Soviets—could hide submarines there.

I wanted to punch him. He just smiled and said calmly, "Sit down." I didn't understand why his demeanor had changed. His aides later told me that I had just been subjected to one of the main initiation rites in the submarine force: the Rickover test. It was named for Adm. Hyman G. Rickover, the father of nuclear submarines, who tested the mettle of officers applying to his program by berating or antagonizing them. Just like Rickover, Thunman liked to see if people had enough spunk to fight back. I guess I passed the test.

On my next trip to the Pentagon, I met with Rear Adm. Leland Kollmorgen, who ran the Office of Naval Research, where my proposal was sitting, unanswered. I walked him through my terrain warfare presentation. "That's it?" he said, seeming amazed that the chief of naval operations would be interested in such a crazy idea. One of his staff members, Gene Silva, reminded him about my grant proposal. For some reason, this angered him, and he told the aide to leave the room.

I was then escorted to a conference room for a larger meeting with Admiral Hayward and several other admirals, who rose to attention when he walked in. "I'm not smart enough to understand the implications of your presentation on underwater terrain warfare," he said to me. Then, looking at the others around the table, he said, "Have any of you ever thought about such a concept?"

One by one, the other admirals said no. When it was Thunman's turn, the discussion took a surprising turn. "Well, sir," he said, "you are aware of the special platforms we have that frequently encounter the bottom."

"I am well aware of those platforms," Hayward said, "but that is different from what Dr. Ballard is talking about, correct?"

"Yes, sir," Thunman said.

I didn't know exactly what the platforms were, but it seemed clear that they were talking about super-secret stuff, probably something to do with very sensitive intelligence operations.

"I would like to have Dr. Ballard signed into those programs," Hayward said. I was as surprised as everyone else in the room. That was certainly a signal that the top dog wanted to bring me into the fold.

The meeting was drawing to a close when Hayward asked a final question. "Dr. Ballard, in your presentation at the War College you mentioned a new underwater exploration technology you have under development. What is the status of that effort?" Before I could respond, Admiral Kollmorgen spoke up, "Admiral, that program is being funded by my office, sir."

At that moment, my *Argo/Jason* system was born. Kollmorgen approved grants of $500,000 a year for four years—two million dollars total to design the system. Funny how things really get done, isn't it? But I knew I would need millions more to test it.

After the meeting, Thunman arranged for the Navy's Deep Submergence Systems group to sign me into the highly classified programs that Hayward had mentioned. I had once used *Alvin* to help the group search for sunken equipment off the East Coast, and I already knew John Howland and George Verd, the captains who ran it.

I also knew that Thunman had asked to use *Alvin* to inspect the remains of U.S.S. *Thresher*, a nuclear-powered attack submarine that had sunk during sea trials off Cape Cod in 1963 with 129 men aboard. Woods Hole had turned him down, saying that the submersible's schedule was booked up.

"What do you mean I can't use *Alvin*? I own it," he had responded angrily. The Navy had built *Alvin* and did still own it, it's true, though the National Science Foundation now covered its operating costs and scientists booked it many months in advance.

I sympathized with Thunman, and I saw an opening to make a deal with him. If he would put up the extra money I needed to test my new systems, I would guarantee him use of the robots either in a national emergency or, if no emergency existed, for one month each year, no matter whether he wanted to use them to photograph something like the *Thresher* site or for intelligence operations against the Soviets.

As I learned more about the classified programs, I could see that submarines had long been on the front lines of our efforts to prevent a nuclear war. Rather than just planning to torpedo surface ships, as they'd done in World War II, the Navy's attack submarines were hanging off the Soviet coast, shadowing Russian subs and monitoring missile tests. Given the oaths I took, I cannot say much more about this, but the book *Blind Man's Bluff* (by Sherry Sontag and my co-author, Christopher Drew) reveals a lot about how submarines were tapping Soviet communications cables and retrieving parts of Soviet test missiles from the seabed.

All this pressure to learn about Soviet military capabilities had come at a cost, and the accidental losses of two nuclear subs—*Thresher* and *Scorpion*—were the most painful ones. On *Thresher*, an electrical part failed and caused the nuclear reactor to shut down, and an ice buildup inside the lines prevented its frantic crew from blowing air into the ballast tanks to propel the sub back to the surface. The loss of *Scorpion,* on its way home from an intelligence mission in 1968, was more mysterious. The Navy postulated that either its battery or one of its own torpedoes had exploded, sending it plunging to the bottom of the Atlantic with its 99 crew members.

The Navy knew where both wrecks were: *Thresher* was about 200 miles east of Massachusetts, but *Scorpion* was a good two-thirds of the way across the Atlantic, several hundred miles southwest of the Azores. The Navy had already examined them both from submersibles in 1969, but Thunman wanted to photograph the wrecks with the more sophisticated cameras we were developing. He was hoping that new images would help experts determine why *Scorpion* had sunk. He also wanted to know if either of the steel containment vessels enclosing the subs' nuclear reactors or the nuclear-tipped torpedoes that *Scorpion* carried had leaked radiation and harmed the environment.

I confirmed that we had the equipment to do a more thorough inspection of both vessels. At some point, I happened to mention that all my life, I'd wanted to search for *Titanic,* and maybe I could piggyback that search onto these expeditions. Thunman was taken aback. "Come on! This is a serious, top-secret operation," he said. "Find the *Titanic*? That's crazy. We don't have the money to do that."

Just think, I said, how fascinated the public would be if we could send a robot inside *Titanic*'s wreckage and show images of it floating down the Grand Staircase.

That's not the Navy's mission, he replied, adding that he couldn't possibly ask his superiors for permission to go look for *Titanic.*

I kept pestering him over the next several months, though, and he finally tried to pacify me. "Look, Bob, you can do whatever you want, but you gotta do it within the time and within the money, and that's it." Only if I fulfilled the assignments on the two sub wrecks could I squeeze in *Titanic.* He also agreed to provide me with an additional three million dollars over five years, bringing the Navy total to five million dollars.

I know it sounds crazy, but I was wondering what Admiral Rickover might think of my ideas, and I asked Thunman if he could set up a meeting. Rickover was now 82 and mostly retired, and when I called to get directions to his office, his assistant answered. Then I heard a screaming voice coming toward the phone like a train out of control, getting louder and louder.

"If your concept of terrain warfare was important to the future of the U.S. Navy," Rickover shouted, "I would have thought of it, and I didn't." At that point, it sounded like his phone went flying across the room and crashed into a wall. I chuckled, realizing that I had just gotten the Rickover treatment from the master himself in my one and only encounter with him.

Throughout this process, Secretary Lehman also was behind me. He had endorsed my underwater terrain warfare idea in a speech titled "Going for the High Ground in the Deep Sea," and in November 1982, he was finally going out with me on that dive in *Alvin*. We planned to do it off St. Croix. On our way there, he told me Admiral Rickover had called him the day before and told him I was crazy.

Lehman is a fun-loving guy, and when I told him that the crews from *Alvin* and the support ship were blowing off steam at the beach, he said, "Let's go." We arrived while a wet T-shirt contest was under way. Then a tall guy from our *Alvin* team rushed over, drunk as a skunk, wearing a big stovepipe hat. When I told him that my guest was the secretary of the Navy, he leaned over and started kissing Lehman, just slobbering all over his face. That was more than John had bargained for, and I about died, though he just smiled and wiped his face.

When we went down in *Alvin* the next day, things couldn't have gone better. We boarded the submersible in the dazzling, sunlit Caribbean waters. It took us three-quarters of an hour to drift down to 4,400 feet. The pilot and I then took Lehman on a lengthy tour of what I described as the undersea battlefield, exploring the towering rock walls and plateaus that represented terrain the submarine force could use.

Throughout the five hours or so we were down, I never stopped talking. I kept mentioning that I'd love to use some of the Navy's time to hunt *Titanic*. I wanted to make sure the secretary added his authority to Thunman's comment suggesting that I could.

I played back to him all that I'd heard him say about his aggressive maritime strategy and Reagan's desire to wage psychological war against the Soviets. The president wants to really play with the Soviets' minds, I said, to make them think we can do far more than we're capable of—and we're capable of a lot. So let me find *Titanic*. I can find that ship with this gear. Give me two weeks, and I'll find it, and then we'll go public. Show videos from the robots roaming through the ballrooms. It will drive the Soviets crazy. They'll think that if we're willing to publicize this capability, imagine what our Navy is doing in secret.

Lehman had to agree, of course, that the logic was compelling. When he could no longer stand my buzzing in his ears, he said, "OK, I'll recommend it to the president that we approve it, but just for two weeks."

Reagan's response, as John relayed it later: "Absolutely, let's do it."

—

I FINALLY HAD THE MONEY to create a Deep Submergence Laboratory at Woods Hole, where we could design and build my robots. I'd been relying on my *Angus* team to do the basic work, but now I could hire engineers who specialized in imaging and remote control systems. The first was Stu Harris, who had a master's degree in electrical engineering from Stanford. He would be *Argo*'s chief designer. Stu brought in Bob Squires, who was familiar with video imaging software, along with Andy Bowen, a top-notch mechanical engineer, and Tom Dettweiler, who had experience with towed vehicles. Before long, I also snatched up Dana Yoerger, an MIT postdoc specializing in underwater robotics. Spiess was also trying to recruit him—but I won.

The only spot available for us on the Woods Hole campus was a prefabricated metal building with light green siding that sat down in a deep hollow. You had to cross a little bridge to get from the parking lot to the front door. Compared with the old shingled buildings scattered around the village and the modern buildings on Woods Hole's upper campus, mine was pretty ho-hum, but at least I had a new home to fill.

I decided to focus first on *Argo,* which was going to succeed *Angus* as my main deep-sea scouting vehicle. It was going to be a significant upgrade.

Although *Angus* could snap thousands of color photos, we had to lift it back aboard the mother ship at the end of the day and process the film to see what it had found. Then the ship had to circle back and send scientists down on *Alvin* the next day to take a look. *Argo,* by contrast, would have two sonar systems and three video cameras that could work well in low light, and it would stream the video up to us as it was recording. That meant that if *Argo* spotted something—a hydrothermal vent, a piece of *Thresher,* or the first sign of *Titanic* perhaps—we'd see it instantly on our video screens. We could hover the ship over the spot and explore what we'd found from every angle, saving huge amounts of time. It could make the difference between success and failure on most expeditions.

I was always trying to push the envelope of what was technologically possible. If my engineers raised questions about whether one of my ideas was feasible, I'd say, "What law of physics does it violate?" But I also counted on them to protect me from going too far. They used to joke about making sure I'd "packed a parachute"—I always seemed ready to jump out of a plane without one, and they had to rush over to make sure I put one on. I generally knew where the line was, but that was how I motivated them to do their best. As the sign on my office door said, "If you don't make dust, you eat dust."

We were right on track in mid-1983 when we got another scare. Like a figure in a Whack-A-Mole game, Jack Grimm popped up again, with just enough funds for another quick search for *Titanic.*

Luckily for me, the weather was bad, and he only had a few days to explore. He remained obsessed with the spot where he thought he'd seen *Titanic*'s propeller, and he came up empty again.

After that, I figured I'd better lock in a timetable with the Navy. Naval planners were becoming more interested in using my systems to gather intelligence, even arranging for me to build a smaller version of *Angus* to help Norway monitor its fjords for Soviet submarine intrusions. Admiral Thunman wanted to test *Argo* at the *Thresher* site in 1984 and survey *Scorpion*'s wreckage in 1985. That was fine with me. I said I'd like to look for *Titanic* after finishing the *Scorpion* work, and Thunman agreed to fund me for a three-week voyage that included both.

When I had Secretary Lehman trapped in *Alvin* and I was selling, selling, selling, I had said, "Just give me two weeks, and I can find *Titanic*." That was always a bit of bluster, and now that the opportunity was right here in front of me, it seemed like an even more difficult goal to meet. Thunman's three weeks included several days of travel time each way plus however long we spent at the *Scorpion* site. That meant that I might not get more than 10 days to look for *Titanic*. I started reaching for my parachute.

Grimm's failures had shown that it wasn't easy to pinpoint where *Titanic* had gone down. The 1912 crew had depended on celestial navigation, and they could easily have misreported their final location. The ocean currents were strong in that part of the North Atlantic, too. I was looking at a search area that could be larger than a hundred square miles, and the fastest way would have been with a sonar system that could survey wide swaths of the ocean—something I didn't have.

I had kept in touch with the French scientists and engineers I'd worked with in the past. Jean-Louis Michel, their top engineer, had been developing a sophisticated sonar system that could be towed behind a ship and scan big chunks of the ocean floor on both sides at once. I knew that this might be my last chance to find *Titanic*, and I decided that it might be better to give up some—even most—of the glory to realize that dream. I flew to Paris and pitched French officials on a joint expedition, telling them to conduct an initial sonar search, and that I'd come in after with *Argo* and its cameras. We all knew this meant that the French would have the best shot at discovering the ship, and I might end up in a secondary role, just photographing the wreckage for the history books. But that seemed like a better choice than risking it all by myself and coming back empty-handed like Grimm.

We shook hands on the deal, and I returned to get *Argo* ready. After all the effort it had taken to secure funding and to build my robot, I was finally putting my latest creation to work. It was going to be the first time I had ever studied a shipwreck in the deep, and it would be a heck of an introduction.

—

THE SHIP'S WINCH LOWERED *ARGO,* a white sled about the size of a small car, into the frothy depths of the North Atlantic. To make sure I didn't miss anything, I wanted to start where there was no sign of *Thresher* and drive toward debris. The Navy had used its bathyscaphe, *Trieste II,* to explore the wreck in the 1960s. As we approached the sub, I envisioned that it would be like crossing a borderline, and suddenly there would be debris everywhere. *Thresher* had been dead in the water when it went down, and the Navy knew that the sub had imploded at about 2,400 feet.

When a submarine implodes, it's a violent event. Everything collapses inward, and then everything explodes outward. I assumed that because there was no forward trajectory, the debris must have fallen pretty much straight down, landing in a circle around the main part of the wreckage. But the actual scene turned out to be not that simple.

The first fragments we saw on *Argo*'s video stream were bits of piping and electrical cable. The wreckage was frightening to behold. It looked as if the mighty vessel had been put through a shredding machine. Fragments lay scattered all over. After about 12 hours of mapping, I began to see a pattern. The pieces closer in toward the main part of the ship were the heaviest.

One image that hit me emotionally was the tail section. It had broken away, and it looked like a giant had crushed it like a toy in his hand and thrown it down onto the ocean bottom. It reminded me of the scene in Homer's *Odyssey,* where Odysseus and his men watch the Cyclops, a one-eyed giant, eat six of their fellow crew members. It was just like the Cyclops had taken *Thresher* and crushed it in a fit of anger. A giant called pressure had devoured them all.

As *Argo* flew over the main pieces, the remains of the sub identified it for us. The sail—the vertical tower that rises above the hull in the middle of the submarine—lay on its side on the ocean floor with *Thresher*'s number, 593, perfectly preserved. The periscope and a diving plane were recognizable, too, and the paint looked brand new, even though the sub had sunk 21 years before.

Nothing in the debris suggested human remains. That did not surprise me. Marine life eats everything but bones, but at such depths, the ocean is undersaturated in calcium carbonate, so bones just dissolve in about seven years. But we did find the containment vessel for the ship's nuclear reactor. It had plunged into the deep clay of the ocean floor, creating a monstrous crater. Only a tip of it was visible, with no indication on the seabed that any radiation had leaked out.

Over the next couple of days, I began to understand that the debris field was not shaped like a circle. Instead, it stretched out in a long line, more like the tail of a comet. My brain went *poof, click,* and principles of physics kicked in. Now I understood what had happened. The heavier objects go straight down. The lighter objects sink at a slower rate, and prevailing currents carry them farther away. In *Thresher*'s case, the debris field stretched for roughly a mile.

It seems like common sense, but it's not the kind of thing you think about until you see it. The key to finding sunken ships was to search for the long debris trail and follow it back to the vessel. This was the great lesson *Thresher* taught me—and one that would soon help me immensely.

CHAPTER 6

MAY GOD BLESS THESE FOUND SOULS

A s the French ship *Le Suroit* approached the *Titanic* search area on July 4, 1985, I was celebrating Independence Day with Secretary Lehman and retired *CBS News* anchor Walter Cronkite on Martha's Vineyard, summer retreat for the rich and famous and their lucky friends. I was, of course, in the latter category, thanks to the friendship I'd struck up with John Lehman since we met in 1981.

Every year since, we'd gathered for the Fourth of July at the home of Spike Karalekas, a Naval Academy graduate and lobbyist who was one of John's closest friends. We'd take in the Edgartown holiday parade and have hot dogs on the lawn. Spike and his wife, Tina, who worked for First Lady Nancy Reagan, always drew an eclectic crowd, from Wall Street traders who didn't want to talk about how they made their money to people like me who could entertain the others with tales of adventure.

Cronkite was a lovely guy, like a father to us. You could see why he'd become the most trusted man in America. He loved science and had covered most of America's space launches. He'd also gone down in *Alvin* with me to see hydrothermal vents. He and Spike were as enthusiastic about my quest for *Titanic* as John was. Early on, we'd dubbed our little group—far too cavalierly—the "Top-Secret Committee to Re-Arrange the Deck Chairs on the *Titanic*." I'd promised to give them progress reports.

Now I was whispering in their ears that the moment was here.

I was finally getting the chance to chase a dream I'd had for so long. I had put together the best team I could to find the biggest prize out there and bring emotional closure to one of the most heartbreaking tragedies of the past century. After all, the *Titanic* story had it all—grand ambition of shipbuilders, arrogance of the captain paying little heed to ice warnings, lack of lifeboats for nearly half the 2,200 people on board. The ship was supposed to be unsinkable.

Just imagine you were there on that calm, moonless night, April 14, 1912, sailing from England to New York on the maiden voyage of the largest and most luxurious ship of its time. *Titanic*'s wireless operators had warned Captain Smith and other officers earlier in the day that some ships had encountered icebergs in their paths. Smith shifted *Titanic*'s course slightly to the south but did not order the crew to slow down. After he had gone to bed, a huge iceberg seemed to rise out of nowhere. The quartermaster quickly spun the wheel that controlled the rudder, avoiding a head-on collision, but the starboard side of the hull scraped along the iceberg below the waterline.

Jack Thayer, a 17-year-old traveling with his father, a wealthy Pennsylvania Railroad executive, and his mother, a Philadelphia socialite, later wrote that in his first-class stateroom, it felt like the ship had been pushed softly. "If I had had brimful *[sic]* glass of water in my hand not a drop would have been spilled, the shock was so slight," he recalled. But travelers in the third-class berths below-decks were thrown from their beds. Seawater started pouring in.

Captain Smith and Thomas Andrews, the ship's designer, hurried down to see what was happening. Andrews counted the compartments that were flooding—one, two, three, four, five. The ship could survive flooding in the first four compartments in the bow, but not the fifth—with that one filled, the weight of the water would tip the boat toward disaster. There was no rejiggering the math. It was absolutely clear. *Titanic* was going to sink, and quickly.

The passengers had no idea yet that the ship was doomed. Some frolicked among chunks of ice that had landed on the bow—until they saw the crew uncovering the lifeboats and firing distress flares. In the chaos that ensued, some passengers struggled into lifeboats while others were stranded belowdecks. The men stoking the coal-fired boilers and running the engines fought feverishly to keep the power on. Through it all, the ship's band played on, first ragtime tunes and then church hymns.

Why do the deaths of more than 1,500 people retain such a hold on us, more than a century later? Walter Lord, who did more than anyone to chronicle the disaster in his book *A Night to Remember,* put it so well, saying that we can see ourselves going through the same emotional stages with *Titanic* passengers and crew members, from disbelief that anything is wrong, through gradual recognition of the danger, and finally to realization that there is no escape. Watching them go through this, Lord wrote, "We wonder what we would do."

Something else had stuck in my head, something I didn't really know what to make of. At some point, Bill Tantum—"Mr. Titanic"— had shown me drawings based on Thayer's eyewitness account. Thayer had jumped overboard right before *Titanic* sank, and he was able to look back and watch the horrible scene. The drawings showed the great ship breaking in two, with the bow sinking before the stern. The jagged ends of the two halves that had ripped apart—the back of the bow and the front of the stern—tilted down as each section went under. But the official inquiry into *Titanic*'s sinking had disregarded that account, concluding that the ship had gone down intact.

I had sent copies of the drawings and most of Bill's other *Titanic* materials to my friend, Jean-Louis Michel, the French engineer who was my co-leader on the expedition. Remember, under the deal I'd made with the French, they were going to find *Titanic* with their powerful new sonar, and I was supposed to come in behind them and photograph the wreckage with my robotic cameras. I'd willingly accepted a secondary role.

As it turned out, the French team didn't even want me there for the first couple weeks of the search. Jean-Louis's colleagues wanted to find *Titanic* all by themselves. They wanted the headlines. "French Find *Titanic*; Ballard's Home Fishing," or something like that. I'd be a footnote.

—

I FLEW DOWN TO MEXICO after the Fourth of July picnic to test an old Navy robot that I was hoping to modernize, figuring we could use it to take videos inside *Scorpion*—and maybe *Titanic*. I almost got stuck there. A Mexican gunboat pulled up to our ship and the officer in charge barked, "Pack your bag, bring your passport." I couldn't very well refuse by explaining I had a secret Navy mission to get to. He hauled me off the boat and ordered me to leave the country within 24 hours. (I later found out it had something to do with a diplomatic spat between the United States and Mexico.) All flights were booked, but I improvised. I crawled through a hole in the airport fence and found a guy fueling his Cessna. After assuring him I was not running drugs, I hitched a ride to Tucson, hoping he wasn't running drugs either.

So began what I consider the luckiest chapter of my life. I usually bristle when people call me lucky. My brother, Richard, always did. When you do something unpredictable, many people have a hard time processing how you pulled it off—and why you were the one who did what others couldn't. They never acknowledge the creative thinking, the willingness to push yourself to the point of exhaustion, and, yes, the often desperate improvisations that went into it. When it came to finding *Titanic,* all of these things came into play—including luck, I readily admit.

I flew to St. Pierre, an island off Newfoundland, on July 22, 1985, with a National Geographic film team, including my friend Emory Kristof, to join the French. They had been out looking for the past two weeks, and now they were stocking up on wine, cheese, and other supplies for the second half of their search, which would last another 17 days. Over dinner, Jean-Louis told me they'd run into bad weather and hadn't found anything. As we headed back out to the search site, I bent over the chart table with him, looking at plots of the paths they'd followed, towing their dark orange sonar vehicle back and forth through the 100-square-mile search box. I could see that high winds and strong currents had pushed them off course on their first pass along the edge of the box, leaving an elongated triangular patch that had not been searched.

The French had done the right thing, however, and kept moving in toward the center of the box, toward the point where we thought we had the highest probability of finding *Titanic.* After I joined them, they kept mowing the lawn, as we call it, with their sonar sweeps, covering 70 to 80 percent of the box, plus an extra area we added outside the initial box. But when they ran out of time and had to return the ship for other uses, they still hadn't found anything.

So now, instead of coming behind the French and just photographing the shipwreck that they had found, it was up to me to find it—and I wasn't sure if I could.

I'd have to map the *Scorpion* site first. Even if I could do that quickly, I'd have much less time at the *Titanic* site than anyone who had searched before. I had great cameras, but the only sonars I had were a small side-scanning one on *Argo,* my newest robot, and the crappy system on R/V *Knorr,* the Navy vessel Woods Hole operated, which wasn't any better than what you'd find on a typical fishing vessel. Neither was designed to conduct a broad, systematic search like with the sonars that Dr. Spiess and the French had used. I'd have to rely on my cameras to find the wreckage, something that had never been done in the deep sea.

—

JEAN-LOUIS AND I FLEW to the Azores to board *Knorr* and head to the top secret *Scorpion* site. We arrived on August 12, and I was glad to see my team and equipment ready to go. There on *Knorr*'s stern were *Argo,* with its black-and-white video cameras, and *Angus,* my old standby, which could take sharp color stills. Cranes had lowered metal vans containing our command center onto the deck, including a lab to develop *Angus*'s 400-foot rolls of Kodak film. Members of my A-team from the Deep Submergence Lab were among the 49 people on board: Stu Harris, my chief engineer and *Argo*'s designer, and veteran *Angus* handlers like Martin Bowen and Earl Young, backed up by my best engineers.

We set sail on August 15. Also on board were Jean-Louis, two other French officials, and Emory Kristof, who had shared my interest in the pursuit of *Titanic* since the failure of my *Seaprobe* expedition in the late 1970s. Three U.S. naval personnel camped out in the control van—a constant reminder that the Navy's investigation was my real mission and that I could only look for *Titanic* after I surveyed *Scorpion*'s wreckage.

As far as many of the crew members knew, we were just heading out to resume the search for *Titanic*. Given the highly classified nature of the mission, the Navy would only let me tell those with a need-to-know status what was really happening. How could I conceal our first stop over *Scorpion*'s wreckage? It was south of the Azores. *Titanic* was west. I was waiting for someone to say, "Bob, why is the sun rising on our port instead of our stern?" We told everyone we were testing equipment for the Navy.

I needed to make sure that *Argo* and *Angus* did work—and quickly.

Knorr's big white crane dropped *Argo* into the water, and our winch operator lowered her more than 11,000 feet, down to the seabed. We began marking the length and width of *Scorpion*'s debris field so we could videotape it. It didn't take long to see how different *Scorpion*'s wreckage was from *Thresher*'s. Although *Thresher* had looked like it had been shredded, the images on our monitors showed that *Scorpion* had broken into three main pieces.

The back part of the sub had imploded and telescoped into itself. The resulting explosions had spit *Scorpion*'s propeller and shaft out more than 100 yards. The steel containment vessel surrounding *Scorpion*'s nuclear reactor had also broken free, as *Thresher*'s had, sinking like a cannonball and burying itself in the clay. Only a little edge was sticking out. *Scorpion*'s forward portion, including the torpedo room, was largely intact. It had flooded, which kept it from imploding.

Navy officials were hoping that our photos would yield more clues about *Scorpion*'s fate, and they wanted us to look for any signs that the Soviets had visited the wreckage. The clay on the ocean floor holds impressions for decades, and we could see imprints from *Trieste II* in 1969, but we saw no sign that the Russians had violated *Scorpion*'s resting place. Good news.

I wanted to do a bang-up job for the Navy, but I was also constantly thinking about how to find *Titanic*. I'd noticed that

Scorpion's debris trail stretched for roughly a mile, just like *Thresher*'s, with the lightest debris extending the farthest out.

In four days we mapped *Scorpion*'s debris field completely, and we got a thumbs-up from the officer in charge of the Navy team. His team snatched our *Scorpion* data, stashed it in a safe, rolled the tumblers, and it was heigh-ho, *Titanic,* here we come.

As *Knorr* turned to the northwest, my mind was buzzing with thoughts of debris trails. *Thresher* had imploded, and so had part of *Scorpion*. The implosions and resulting explosions had spewed out torrents of shattered steel mixed in with much lighter parts. If *Titanic* had sunk in one piece, water would have rushed in through the openings on the deck. With water pressure equalized inside and out, there might not have been any implosions. That meant there might not be much of a debris trail.

But what if the sketches based on Jack Thayer's account were right? What if *Titanic* had split open, the two halves plummeting, broken ends first? Then some of what was inside would have tumbled into the ocean like salt and pepper pouring out of shakers.

I visualized the ship not just breaking in two but falling to the bottom, with the heaviest pieces heading straight down and lighter ones drifting in the current, just as I had seen with *Scorpion* and *Thresher*. It played out like a film in my head and, all of a sudden, it was as clear as a bell. I shouldn't be searching for the ship. I should be searching for the debris trail. A mile-long trail would be easier for me to find than an 883-foot ship that was maybe in one piece, maybe not.

We arrived at the *Titanic* search area on August 24. First we checked an underwater canyon nearby, because I figured that *Titanic*'s debris might have flowed down into the canyon's main channel and remained hidden from previous search teams. We sent *Argo* out to look at some of the earlier sonar targets, but there was nothing at Grimm's propeller site. Other objects—which we dubbed "Ryan's Madness" and "Spiess's Obsession," given how much time they had spent looking at them—turned out to be rocks. Now, with nine days to go before *Knorr* would have to head home, I was excited to test my debris trail strategy.

I decided to tow *Argo* back and forth to the east of where the French had worked, in case *Titanic* had been traveling more slowly than reported. I set the lines of *Argo*'s pathway roughly a mile apart, running east to west. That way, if the currents had carried *Titanic*'s debris in a mile-long trail from north to south, we might be able to intersect it as we worked our way north.

Thank goodness, the seas were calm, and it was easier for us to keep our search vehicle on course than it had been for the French. But we still had to deal with those pesky laws of physics. *Argo* was dangling at the bottom of a cable that was almost two and a half miles long, and if our ship moved at more than a crawl, friction with the water caused the sled to lift up too high for us to see anything. Martin Bowen was flying *Argo* from the console to my left. We had reached the far end of one line and were ready to turn the corner and start mowing down the next in the opposite direction when suddenly Martin let out a yelp.

I looked over at him from the plotting table and instantly realized that we had a serious problem. We'd slowed the ship, causing *Argo* to fall toward the bottom as the drag on the cable dropped. To keep it from hitting the seabed, Martin began reeling in the cable. But the take-up drum on *Knorr* wasn't moving fast enough, and the extra cable fell off and got tangled in the gear, shredding its protective armor. I rushed out with the crew, and we rigged up a splint and recovered my multimillion-dollar baby. But would we be able to lower *Argo* into the depths again or get any video from it?

Fortunately, the core of the cable that transmitted the video was not damaged, and the damage to the outer sheath was high enough on the cable that it would not affect us as long as we didn't have to search in any deeper water. Sheer luck: We were heading into an area that was not quite as deep.

The next three days, we made monotonous runs. So much of deep-sea exploration is just plodding along, seeing little more than fuzzy, black-and-white images of mud. The crew on watch squinted and rubbed their eyes, struggling to stay focused. The heady expectations, the sense of exhilaration that we were about to do something monumental, were fading fast. It was tough on morale, as each four-hour shift gave way to the next with nothing new to report. Then the tensions boiled over.

Argo's small sonar picked up an image of something off to the side. It seemed as big as *Titanic*. Jean-Louis and I were sure it was a natural phenomenon, another wisp like the ones Grimm's team had been chasing. But Emory Kristof, in front of everyone in the van, insisted that we swing back to take a look. Others supported him.

It stung that a good friend like Emory would challenge my judgment publicly, and I didn't want to lose the crew. One principle guiding me on this search was that we had to be disciplined. As Dana Yoerger, my robotics expert, recalls, my instruction to the crew had been: "Once you make a plan, unless you see something that says T-I-T-A-N-I-C on it, you're not stopping for anything."

I turned to Jean-Louis. "Come out on the deck," I said. "Let's talk."

I had to remind myself that this was the first time I'd gone after a shipwreck whose resting place was unknown, and my debris trail theory was still just that, a theory. Jean-Louis and I decided to compromise with Emory, and I went back in to tell the navigator. Emory came up to ask what I was going to do. "Ask the navigator," I snapped and kept on walking. I was still mad that he made me veer from my course. As Emory soon found out, we had decided to swing back past the area that interested him, and we all saw on the monitors exactly what Jean-Louis and I had expected to see: sand dunes—giant ones, but still sand dunes. That ended that argument.

I pride myself on being an optimist, but a couple of evenings later I was still feeling pretty low, wondering if we could pull this off. When the watch changed at midnight on September 1, we had just over four more days to go. As chief scientist, I like to be in the control van every four hours, to make sure the new watch team knows what the old watch team did. We were about to pass through that triangle in the original search box that the French had missed. Everything seemed to be going fine, so I headed up to my state-room. I remember hearing Marvin Gaye singing "I Heard It Through the Grapevine" as I left the van.

My cabin was next to the captain's, up in the clouds at the top of the ship. It was away from the action, a good retreat. Even as a boy, I'd been able to fall asleep on my way to the pillow. I'd also gotten good at squeezing in a couple of deeper cycles of rapid eye movement between watch changes at sea. But tonight was differ-ent. I was restless. I was reading Chuck Yeager's autobiography, and remembering how my dad had flown with the celebrated test pilot over the Mojave Desert. Then someone knocked on my door. A voice called out. It was the cook. *Why,* I thought, *is the cook here at one in the morning?*

"The guys think you should come down to the van," he said.

He didn't have to finish his sentence. The watch team must have seen something big enough that none of them wanted to leave their screens. I could just hear the conversation. *Somebody had better get Bob, but I'm not the one who's going.* So they told me through the grapevine.

I hopped back into my blue jumpsuit, and *bam,* I was out of there, gliding down two decks with my hands on the railings—turning a corner, turning a corner—my feet barely touching the steps. I blasted into the command center, looking toward the screens on the front wall. I saw debris on the three screens displaying *Argo*'s black-and-white video feeds, and I was looking at the plot to see where we were. There was a compass heading, a bearing. I could see how high off the ocean floor we were. I was sucking it all in, forming a mental image.

Then I heard Stu Harris's voice telling me what they'd just seen. It was a ship's boiler. The crew members were buzzing with joy, and one of them rewound the tape so I could see it for myself. The tape showed that Stu had been the first to spot some smaller debris. You could hear him say simply, "There's something." Moments later, Bill Lange, whose job was to document what we saw, shouted, "Wreckage!" That was followed minutes later by Stu's "Bingo!" and then Bill's "It's a boiler!"

The video images confirmed their enthusiasm. There it was, a metal cylinder, more than 15 feet in diameter, with rivets and three large doors for shoveling in coal—just like the boilers in pictures we had of *Titanic*.

"Goddamn," I said. "Goddamn."

There it was, one of 29 boilers that had created steam for *Titanic*'s engines. It was a signature piece. We'd found the debris field, about 12,500 feet down. Bull's-eye!

I looked at my friend Jean-Louis. He knew what I was thinking. "It was not luck," he said. "We earned it."

As we passed over more of the wreckage, everyone celebrated with cups of Portuguese wine. It made me nervous. Some of the crew seemed to be losing focus, and I feared that the wreckage might snag the equipment. I ordered *Argo* brought up to almost 200 feet to avoid the debris below. I didn't want to lose our robot eyes the very moment we had found *Titanic*.

Around 2 a.m., someone remarked that we were approaching the time of night when *Titanic* had sunk into a sea as calm as the one we had now. It wasn't until this point that the emotion of the tragedy fully hit me. I know this sounds odd, but it was quite unexpected. I had never been a *Titanic* groupie. Sure, I'd wanted to find it, and I'd been very competitive about that. But a world tragedy had played itself out on this spot, and now the site itself took hold of me. Its emotion filled me and never let go.

I said I was going out on the fantail for a few moments of silent reflection for those who had lost their lives here. Most of our team members came up with me. Seventy-three years before, the waters around us were teeming with people crying for help and frantically trying to reach lifeboats. Jack Thayer, the teenager who was in the water with them, described the scene in words I'll never forget: "Then an individual call for help, from here, from there; gradually swelling into a composite volume of one long continuous wailing chant, from the 1,500 in the water all around us. It sounded like locusts on a midsummer night, in the woods in Pennsylvania."

As Thayer recalled, "This terrible continuing crying lasted for 20 or 30 minutes, gradually dying away . . ."

The sea had claimed all these people. Some drowned inside the ship as it fell beneath the waves. The bodies of others who ended up in the water without life jackets just rained down to the ocean bottom. When you drown in the deep sea or die of hypothermia, you don't float to the surface again. You sink, and the pressure of the deep keeps you down. The cold tentacles of death grab hold and don't let go.

I wanted to share the big news of our find with John Steele, the Woods Hole director, even though he hadn't supported our search. It was the Sunday of Labor Day weekend. I waited until a reasonable hour and called the Woods Hole operator, asking to connect with Steele at home. The operator came back with his answer: He was too busy to take my call.

—

WE HAD FOUND DEBRIS, but not the ship, and we needed clear pictures of a sizable part of *Titanic* to convince the world we'd found it. The boiler—yeah, we had that. But if all we had was a picture of a boiler, well, who says that's really it? I knew we were close, but we weren't over the finish line. If the ship had split apart, where was the bow? Where was the stern?

Argo's video cameras picked up pieces of hull plating and other debris that clearly had come from a large ship. As we repositioned our navigational transponders around the debris field for better tracking, *Knorr*'s bottom-sounding sonar made contact with a 100-foot-tall object to our north. It was 13.5 miles southeast of *Titanic*'s last reported position, but given its proximity to the other debris, it seemed likely to be a significant chunk of *Titanic*'s hull. How funny, I thought, that *Knorr*'s crappy Fathometer—not the more sophisticated sonar systems Ryan, Spiess, and the French deployed—found what seemed to be a large part of the ship. Looking for the debris trail had taken us to the right area, and now it looked like we had just kind of run over it.

Everyone agreed that it had to be a big piece of *Titanic*. We stood at the chart table, and Jean-Louis superimposed his original search map over this spot.

"Merde," he said.

He had missed *Titanic* by less than 3,300 feet on his first sonar run.

Talk about bad luck. If the currents had not pushed his ship and the sonar it towed off on an angle, he'd have run into the wreckage and won the acclaim for finding *Titanic*. The French were so close, in fact, that I'll bet they picked up some signs of small debris. But on a sonar screen, it would have looked just like rocks, nothing of interest.

▲ *Here am I, on the left, with my brother, Richard, and our pony, Bucky, outside our home in Gardena, California, in 1946.*

◄ *My brother and I hold hands with our sister, Nancy Ann, as we look out at the Pacific Ocean. At that time, in 1948, our house was just 100 yards from the beach.*

Christmas Eve, 1954. Three Ballard kids—Richard, Bob, and Nancy Ann, left to right—under the decorated tree, and then the whole family, including my parents, at the dinner table.

◀ *I played split end and defensive halfback for my high school's varsity football team, the Downey Vikings. Here, late summer 1959, I'm ready for action, but my football career ended when I got a brain concussion the second game of the season.*

▼ *Commissioned as a second lieutenant in Army Intelligence as I graduated from the University of California, Santa Barbara, in January 1965, I'm proudly saluting in front of the family car.*

For my job as a dolphin trainer at Hawaii's Sea Life Park, 1965–66, I had two responsibilities. Some days we put on a show (above), but other days we conducted research on dolphin behavior (opposite). It was a great job for someone who loved being in and around the ocean.

Margie and I were married in California, on July 1, 1966 (opposite). By 1971, after moving to the East Coast, we had two little boys (above): Todd, the happy little blond, and Dougie, then still a toddler.

We spent time every summer in Montana at Margie's father's cabin on Whitefish Lake. Here we are in the summer of 1974, posing with her dad's pickup.

▲ Dougie (left) and Todd play in the backyard with our dog Tassie at our home in Hatchville, Massachusetts, on Cape Cod, 1979.

◄ Todd and Dougie spent hours practicing hockey shots in our basement, and Todd made the Falmouth High junior varsity team in his sophomore year, 1984.

Every one of my kids has gone to sea with me. Todd went on my Bismarck *expeditions; here we're together at the press conference afterward (above, top). Dougie, working here in the conservation lab, joined a Black Sea expedition in 2003 (above). Ben (opposite, top) joined several* Nautilus *expeditions and impressed me with his knowledge of ancient history. Emily (opposite, below) is a* Nautilus *crew member; here, she's conducting tours of our ship in Fort Lauderdale.*

◄ *Born in Kansas yet a California kid, I still found myself right at home with the Boston Sea Rovers during our annual lobster dives.*

▼ Alvin *takes a selfie, 1976: Exploring the Cayman Trough 12,000 feet underwater, the submersible would position a camera on the ocean floor and signal it to capture a deep-sea self-portrait.*

▲ *Investigating the Galápagos Rift in 1977, we discovered new life-forms such as these rose-colored tube worms thriving in the warm water coming out of hydrothermal vents. The discovery revealed a new ecosystem based on chemosynthesis—and rewrote the biology books.*

▼ *Exploring the East Pacific Rise off Mexico in 1979, we discovered tall stacks of mineral deposits now called black smokers. Formed by superhot fluids from active magma chambers below, they were so hot they could have cracked* Alvin's *viewports had we moved in too close.*

To explore the Reykjanes Ridge south of Iceland, we towed the nuclear-powered submarine NR-1 from the U.S. Navy base at Holy Loch, Scotland (at left). In those days, I took photographs through the sub's viewports (below).

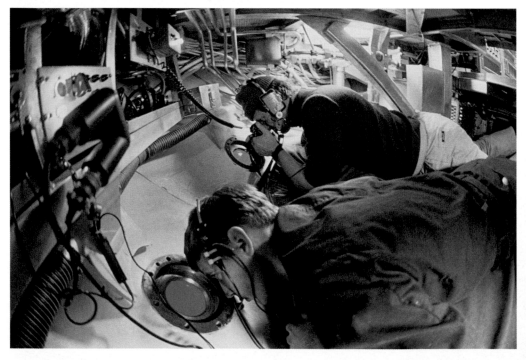

▲ *Aboard R/V* Knorr, *my French colleague Jean-Louis Michel and I stand speechless at the moment of discovery of* Titanic, *September 1, 1985.*

▼ *The telltale shape of a boiler, with rivets and doors for shoving in coal, clinched the identity of* Titanic.

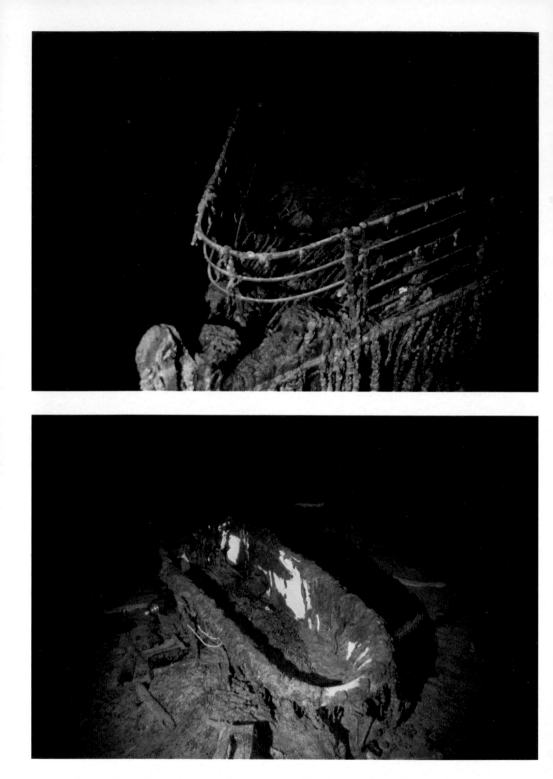

Returning to Titanic *a year later, 1986, with* Alvin *and JJ, along with much better cameras, we photographed every inch of the wreck, including the bow (top) and poignant personal artifacts, like this bathtub.*

We did this together, I told Jean-Louis, trying to console him. "I had my chance," he said, "Paris will look at it as my failure." Then, stand-up guy that he is, he quietly turned back to the table to help plan our next *Argo* runs. I really felt for him, but I also appreciated that. We had a lot to do, and the fickle North Atlantic weather was turning dicey.

As the winds picked up on September 2, we prepared to make the first scouting run over what we felt sure was the main part of the ship.

If you look at pictures of *Titanic* in all her finery, you'll see a cobweb of wires above her. Telegraph wires, looking like trapeze lines, ran high along the length of the ship. Guy wires slanted up from the deck to stabilize the funnels, or smokestacks. If she was upright—and we didn't know at this point—*Argo* might have to descend into the cobweb to get good pictures. I was terrified that if I got too close, the lines would grab *Argo*, or we would smash into a funnel, and I'd lose the ball game.

I stood in front of the bevy of monitors in the control van, taking in all the data and getting into my zone. I was totally inside myself, visualizing it, trying to create a three-dimensional picture in my mind of what was down there. It was like I was doing a mating dance with *Titanic*.

Everyone was gathering in the van, wondering when I was going for it. *We're running out of time,* I told myself. *I've got to go for it.*

We position *Knorr* on the starboard side of the wreckage and come to a halt several hundred feet away, pointing right at the large piece we'd discovered. Earl Young has his hand on the joystick, holding *Argo* safely at 50 meters—164 feet—above the seabed, ready to lower or raise it at a moment's notice. I say to move *Knorr* as slowly as possible. I just want to creep over there.

We're all looking at the video monitors, and it's like we become *Argo*. We're too high to see anything. I'm waiting for *Argo*'s

Fathometer, which looks down and slightly forward, to pick up signs of the wreckage. All of a sudden, our altitude reading drops in half, to 25 meters, indicating something large rising off the ocean floor. We're about to pass over *Titanic*.

It's a scary moment. On Earl's monitor, the ship seems to be rising up at us. He's looking at an electronic image with two horizontal lines. The one showing *Argo*'s altitude stays in the same place, but the line showing the bottom depth is rising rapidly. It looks like they're going to collide, and all our instincts are to raise *Argo*. I'm sitting next to Earl, right up against his ear, like Braveheart as the goddamn Brits were charging them, saying, "Hold." He wants to panic. He wants to bring up the vehicle. I'm thinking, *If he does, we won't see anything.* "Hold."

In fact, I order him to drop down another 16 feet. Now we're only 66 feet above the wreckage, just under the top of the funnels if some are still standing beyond the range of our lights. Earl has a grip on the lever. He's almost squeezing the pulp out of it. His hand is going white. Then, slowly, faint and fuzzy on the black-and-white video stream, *Titanic*'s hull comes into view. "She's upright," I say in my best deadpan command voice. "All stop."

As *Knorr*'s forward momentum slows, I'm trying to figure out what we're looking at. I'm seeing the very edge of the ship and then, suddenly, the boat deck. We're inching toward the middle of the starboard side, where some of the lifeboats were launched.

Now all the images of *Titanic*'s floor plan are racing through my head. What's next? The frigging funnel! That's really high, and I'm imagining *Argo*'s going to hit that thing. But no funnel there. Instead, a big hole. Thank God.

Knorr stalls, centered perfectly over the funnel opening. We're dead in the water, and God's in charge. *Argo* begins to rotate a little on its cable, and I just let it drift. By sheer luck, we drift down the center axis of the ship toward the bow. I couldn't have choreographed it any better.

Everyone was silent, glued to the action. The command center was full of people from all watches, plastered against the back wall. No one wanted to miss this.

Next up, *Titanic*'s bridge. But there's no bridge. All we could see is the base of it, like the foundation of a house that's gone. Then we see the toppled foremast, with the crow's nest, where the watchman had first spied the iceberg. We keep drifting forward, and *Titanic* angles down as we approach the cranes that load things into the hatch. Folded inward, they look like a lobster with its arms crossed. Then we see the hull, and we know the raised deck with the crew quarters is coming.

We were just drifting slowly. *Titanic* was unveiling herself to us. No one was blinking. Our eyes were drying out, because we were not going to close them for a nanosecond. It was an "Oh, my God!" kind of moment. *Argo* started to drift a little to starboard. We were going over those monstrous bollards and capstans and the chain links of the anchor, so huge each link weighs 175 pounds. Then we went out over the railing and into nothing, just the blackness of the sea again.

At that point, everyone exploded. We'd done it. That little tour must have taken just under six minutes, and we had not run into any of the cobwebs. Everyone in the control van went crazy, yelling and hugging each other.

Jean-Louis and I stood quietly in the middle of it all. We were on the verge of tears, just taking it all in and marveling at the achievement. I'd been chasing this dream, hoping for this moment for years, and now it was finally here.

But we still had a lot to do, and the clock was ticking. We had fewer than three days left aboard *Knorr*. Later that afternoon, we took *Argo* farther aft, trying to figure out how much of the ship was still there. We were feeling it out from a safe height, getting to know the beast. Could we get even closer?

We could see a hole above the Grand Staircase, but we couldn't see down into it. There also was a hole where the second funnel used to be. As *Argo* continued aft, the deck sloped down into a vicious scene of twisted metal. We had surveyed about 470 of *Titanic's* 883-foot length. Where were the third and fourth funnels? Where was the rest of the ship?

Jack Thayer was right. *Titanic* had broken in two at its weakest point, between the second and third funnels—a large area with little supporting structure—and the bow and stern had sunk separately. But where was the stern? We wanted to explore further, but bad weather was closing in on us, and we had to reel in *Argo*.

It was almost 11:30 at night on September 2, and I'd been up for 24 hours. One of the things you learn from the sea is patience, because you are not in control. This time, my patience lasted an additional 10 hours. The sea was churning, but I thought we could drag *Angus,* my rugged old dope on a rope, through it. I needed to go home with some of *Angus's* high-quality color still images, not just *Argo's* black-and-white video. We needed the icing on the cake.

With less than two days to go, we lowered *Angus* into the frothing sea. First, we sent it out over the parts of the debris field with the smaller objects. That was the safest bet. If we ended up going too close over the ship, we'd at least have all the other pictures. And there was amazing stuff in the debris field, personal things like teacups, wine bottles, a silver platter, and items from people's staterooms—poignant reminders of those on board.

We didn't have a live link to *Angus,* so we were flying blind. We wouldn't see the pictures until we developed the film. From late afternoon into the evening on September 4, we tugged *Angus* over *Titanic's* bow five times, cameras snapping away. I was still being cautious, keeping *Angus* 20 to 30 feet above *Titanic*'s decks. I knew that the pictures were likely to be fuzzy.

After midnight on September 5, word came from our onboard lab that the photos weren't clear enough. The time had come for one final epic run with *Angus*. I had barely slept in days, and I was exhausted. I also had slipped on the wet deck—I don't even remember how—and injured my leg, and I was having trouble standing. We had until 7:30 a.m. before we had to take up the transponders and get ready to head home. I was going to pull out all the stops, even if it meant damaging *Angus* in the process.

High winds and seas were battering *Knorr*. The launch team had to put on foul-weather gear, and Earl had to tie himself to the ship to turn on *Angus*'s strobe lights and cameras and to drop it into the sea. Adrenaline kills pain, and that helped with my leg, but I was still so worn out that I crawled under the chart table to lie down. I took catnaps as *Angus* made the descent to *Titanic*'s grave. Earl and others prepared everything to get the evocative still-life portraits that she deserved. I figured *Angus* would need to get within 12 or 13 feet of *Titanic*'s deck for sharp images. Even that close, the reds would dissipate, and the photos would be a ghostly mix of shades of blue. But at least they would be clear.

I hobbled over and pulled up a chair right behind Earl. Leaning over his shoulder, I gave the order to take *Angus* down to 13 feet.

"Four meters?" Earl asked.

"Four meters," I repeated.

For three hours, Earl led *Angus* on one blind pass after another over *Titanic*'s bow as the rolling waves above pulled the cable—and *Angus*—up and down. Finally, word came that we had to break off if *Knorr* was going to keep its schedule. We pulled up *Angus*, and then Martin Bowen spent the next several hours in the photo lab. He came out smiling, announcing triumphantly that we had terrific pictures.

Argo's video cameras had also run for most of the first week, and we had shot more than 20,000 frames in 8,000 locations. We had all that to take home and study. It wasn't until we got into the analysis that we realized we'd passed over *Titanic*'s stern. It lay 2,000 feet south of the bow, near where we'd spotted that first boiler.

Dr. Steele, Woods Hole's director, had finally gotten excited about what we had accomplished, and Woods Hole had arranged interviews for me with one reporter after another, using *Knorr*'s ship-to-shore hookup. Just as I started talking to Tom Brokaw, the *NBC News* anchor, I looked out and saw that we were actually sailing away from the *Titanic* site. *Whoa, whoa, wait a minute,* I thought. *I didn't get to say goodbye.*

I had started this quest for its own sake, focused mainly on the challenge. But now I had a deep emotional connection to *Titanic*. Its resting place is one of those places that speaks to you, like Gettysburg or Normandy. It moves something deep inside you because it's seen so much loss of life. I was thinking I hadn't been respectful enough. I hurriedly got off the phone and ran out to the fantail. I needed one more calm moment to say farewell.

On our voyage home, I jotted down some thoughts to say to those on land curious to hear the details of our discovery. *Titanic* lies "on a gently sloping alpine-like countryside overlooking a small canyon," I wrote. "It is a quiet and peaceful and fitting place for the remains of this greatest of sea tragedies to rest. May it forever remain that way, and may God bless these found souls."

CHAPTER 7

TAKING STOCK AFTER *TITANIC*

H elicopters buzzed overhead and small craft swarmed out to greet *Knorr* as we sailed into Woods Hole on September 9, 1985. VIPs and family members, including Margie, Todd, and Dougie, were ferried out for a private welcome. Showing off a bit, the captain blared *Knorr*'s horn and did a full 360-degree turn for the hundreds of people peering up at us excitedly from the dock—Woods Hole folks, mainly, but also a startling crush of media from as far away as London and Paris. Our fans were waving homemade signs and tossing confetti and balloons into the air.

"How do you feel about finding *Titanic?*" the reporters shouted at me.

"I'm glad it's over," I yelled back.

That'll give you some idea of how unprepared I was for what was to come.

I'd had my share of media attention and done the talk show circuit after I'd found the hydrothermal vents and black smokers.

I could handle myself in that world. But this was like nothing I'd experienced. It started with Jean-Louis Michel and me both saying a few words in the Woods Hole auditorium, praising our countries' cooperation and calling for peace for *Titanic*'s lost souls. Then we shook hands and rushed into waiting cars. Reporters were tossing pieces of paper with their names and phone numbers through the window, grasping for exclusive interviews.

But I wasn't saying much more. I was due to fly down to Washington, D.C., for the main press conference, which National Geographic had agreed to host, and I needed to keep my mouth shut until then.

You'd have thought our voyage back to Woods Hole would have been a time of celebration, but it was actually fairly tense. A lot was simmering behind the scenes that I hoped the reporters wouldn't find out about. Jean-Louis and I had been partners in the search, and the French had come so close to finding *Titanic* that I wanted to make sure they shared the glory. On our way back to Woods Hole, we'd sent a small batch of photos to shore by helicopter, carried by one American and one French naval officer, to represent the partnership. I'd made a handshake deal that we'd wait to release our photos in the United States until theirs had arrived in Paris, so the announcement of our discovery could come in a simultaneous press release from both locations. Unfortunately, John Steele, my boss at Woods Hole, had buckled under pressure from U.S. news outlets and let them broadcast the images early. The French were outraged, and so was I.

Steele had never supported my search for *Titanic*. In fact, he'd done everything he could to stop me. He was a purist, preferring projects of solely academic interest. Now, seizing control over a project he had barely approved, he was betraying the great working relationship I'd developed with the French.

I'd also gotten a tip from someone at Woods Hole that Steele was planning to have his security officials seize all the *Titanic* photos and videos, and I was afraid he might dump a lot more of them to the media and turn everything into a circus. As the leader of the expedition, I had the sole right under federal law to control their release. Steele had already shown he didn't care about that, so I donned my other hat as commander in the Navy Reserve and classified the remaining *Titanic* material "Top Secret." I cited Navy sensitivities about wrecks that might hold human remains. I asked the Navy officer who'd commanded our *Scorpion* survey to change the combinations to the safes where the photos and videos were stored—and not tell me the new combinations. That would keep the *Titanic* material out of reach until calmer legal heads could prevail.

The tensions continued as I reached D.C. The French government went to court to try to block the news conference, and an American judge rejected that request in the middle of the night. Still hoping to repair the damage with my French colleagues, I insisted on showing only the few images that Jean-Louis and I had agreed on, so National Geographic assigned an artist to work through the night to add some renderings to the mix. Here I was, at the defining moment of my life, and so much—my relationships with my boss and my longtime exploration partners—seemed to be dissolving in acrimony.

John Lehman came to the news conference and sat right in the front row. Having the interest and support of the secretary of the Navy buoyed me. I kept my presentation simple. The story told itself. The reporters were transfixed by the images—*Titanic's* deck, its anchor chain, the dishes and wine bottles. I wanted to keep them from sniffing around too much, in case they'd heard about the French lawsuit, and so I fed them something else enticing for their stories: a controversy that dated back to days right after *Titanic* sank. It turned out that a British steamship, *Californian,* was under way toward Boston ahead of *Titanic* when it encountered the ice field and stopped. It was at least five to seven miles—and maybe as many as 21—north of *Titanic*. It had sent *Titanic* a warning about the iceberg, but when *Californian*'s officers later saw lights and flares, they thought *Titanic*'s crew was simply entertaining its passengers. The radio operator aboard *Californian* had gone to bed by then, and if the captain had awakened him and told him to put on his headset, he would have heard *Titanic*'s operator calling for help.

Could *Californian* have rescued some of *Titanic*'s passengers and crew members from the icy waters if it had responded to the distress calls? I raised that question once again, to get the reporters' attention. Now that we knew where *Titanic* went down, I said, we could see that *Californian* had indeed been close enough to rescue some of the passengers. The media always crave a controversy, so I gave them one that was 73 years old, and they seized it. By the end of the briefing, I was back in control.

But the media blitz didn't stop. I went on the *Today Show,* the *Tomorrow Show,* and the *Day After Tomorrow Show*. Once I got a chance to catch my breath, I could see what everyone else who's ever been thrust into the spotlight learns: For better or worse, my life had changed forever. I'd been naive—or just weary from lack of sleep—in thinking that finding *Titanic* was the end of the story. That was the scientist in me: Make a discovery and write it up. But the paradigm shifter and promoter in me knew better. Now there would always be some people who wanted to take me down a notch, but others I'd never met welcomed me into their company.

—

STEELE AND I HAD BARELY tolerated each other before, and now I could see we were heading into a full-pitched battle. He and other scientists at Woods Hole had such mixed emotions about what I'd been doing, and now the charge of being a popularizer would be hurled at me again. I'd always felt that Carl Sagan, the Cornell astronomer, had been punished for popularizing science with his *Cosmos* television series when he was rejected for membership in the National Academy of Sciences. I worried the same thing would happen to me. I'd been senior author on more peer-reviewed articles than the vast majority of tenured faculty members at Woods Hole. I certainly didn't want fame from the *Titanic* discovery to shadow my serious scientific work.

On the other hand, it didn't take long to see the tremendous opportunities my new status would provide. Crates of letters were arriving at Woods Hole, and the phone was ringing off the hook. Our quiet little institution had become internationally famous, and pretty much everyone there was irritated with me about that. All they could do was watch as my desk disappeared under a mountain of more than 16,000 letters, many from schoolchildren saying our *Titanic* discovery had made them want to become scientists and undersea explorers, too.

One day, my secretary got a phone call from someone who said the White House was calling. President Reagan wished to invite me to a dinner in honor of Prince Charles and Princess Diana. Thinking it was a prank, my secretary asked them to send the invitation. Sure enough, a huge, embossed envelope soon arrived. It looked like I'd just won an Academy Award. The dinner was to be held not in the grand salons where state occasions normally occur, but upstairs in the private family quarters. I had never been to the White House. I had no idea what was expected of me.

Fortunately, I had someone I could call: Shirley Temple Black, the child superstar who had grown up to serve as an ambassador and chief of protocol for the United States. I had gotten to know Shirley because her husband, Charlie Black, was on Woods Hole's board of trustees. We had had some good times together. I vividly remember belting out "On the Good Ship Lollipop," one of her childhood hits, while driving home from a restaurant together. Shirley was a bit of a prankster. One time she invited me to a fundraising event and asked me to bring a giant tube worm we'd collected at the Galápagos Rift hydrothermal vents so she could put it on ice on the hors d'oeuvres table. The tube worm was the star of the party. I stayed at their home on several occasions and loved to look into her aquarium, which included a diver with air bubbles coming out of its head. Months later I received a small box. Inside was that very diver, standing on a wooden pedestal.

So I called Shirley to get some advice on what to expect and how to behave. Her answer was succinct: Get a room across the street from the White House at the Hay-Adams Hotel. Arrive a few minutes early for dinner and stand to the side and listen as the guests are announced to the press corps.

Margie and I checked into the Hay-Adams. It was so close to the White House that we planned to walk over to dinner. But I was unsure which entrance to use. I took a brisk jog that afternoon, swinging by the White House, and asked one of the guards which door tonight's dinner guests would be using. His trembling hand moved toward his sidearm. It didn't take Army training to figure out he was unnerved by a sweaty stranger, even though I told him I was one of the guests. "Clearly I have made a mistake," I quickly said. I told him I was going to walk slowly backward to my hotel across the street, and that's what I did.

That night, we took a limousine.

The dinner party, about 80 guests in all, was led upstairs, where the president himself greeted us. As I approached him, an aide whispered my name in his ear. He was cordial, but I could tell he didn't really recognize who I was. I proceeded down the receiving line to Prince Charles, who took my hand. Suddenly, the lightbulb went off for President Reagan, and he started congratulating me for finding *Titanic* while I stood there holding the prince's hand. That moment lasted an uncomfortably long time. Next, Nancy Reagan introduced me to Princess Diana. To my surprise, she was extremely shy. She darted out her hand and quickly yanked it back.

The room was filled with famous faces: Alan Shepard, Mikhail Baryshnikov, John Travolta, Beverly Sills, Clint Eastwood, William F. Buckley, Jr. At one point, Shepard was telling a story at his table about balloon riding, and he punctuated it by making a loud *schuss*ing sound. It came out more like the kind of sound you'd never want to make in polite company. The room fell silent. Then President Reagan stood up with a smile and said, "It wasn't me!" His friendly comment broke the ice, and everyone began to relax.

There were toasts, lobster mousse with Maryland crab, glazed chicken, and peach sorbet. Then we were led downstairs, where Neil Diamond performed. People began to dance. When the tune shifted to "Saturday Night Fever," a suddenly un-shy Lady Diana marched boldly over to Travolta, grabbed his hand, and led him onto the dance floor. He seemed surprised and uncomfortable. But not Lady Di, who knew how to take over a room. Everyone stopped to watch. Diana was in a form-fitting, floor-length, midnight blue velvet dress, and she kept up with Travolta through a double spin. We all held our breaths, afraid she'd tumble to the ground, but she never missed a beat.

Prince Charles, who had been watching a bit grumpily from the sidelines, chose that moment to stalk onto the dance floor with the wife of the British ambassador, easing his way over to his wife and seeming to signal her that enough was enough. She ignored him and kept on dancing, and the royal couple didn't leave till past midnight. Margie and I then took the limousine back across the street to the hotel. I later got a $900 bill for the two-block round-trip.

The public was entranced with all things *Titanic,* and many other invitations followed. Not long after the White House dinner, I was asked to give lectures on *Titanic* aboard *Queen Elizabeth 2* on a trip from England. They gave us two first-class cabins, and Margie, Dougie, and one of his friends came along. The odd thing was that we had to be diverted from New York to Baltimore. The reason? A warning about icebergs.

It was rapidly dawning on me by that point: *Titanic* was not just another trophy. It had given me sufficient status to be included in such company, and thus greater power to implement new projects I was dreaming about. I had been so moved by those letters from thousands of students that I wanted to set up an educational program in science and engineering. Given how well preserved *Titanic* was, I'd also begun wondering what the state of preservation might be for older, wooden ships on the seafloor: British schooners, Spanish galleons, even vessels from the classical age. I proposed to Steele that I create a Center for Marine Exploration at Woods Hole to run these and other projects. It was time to take stock. My mother was right. I didn't want to go down in history as the guy who found that rusty old boat. I was 43 years old and had a lot of years ahead of me. Now that I had the world's attention, it was time to decide what was next.

—

THE NEXT SUMMER I headed back to the North Atlantic for a second visit to *Titanic* aboard Woods Hole's R/V *Atlantis II,* the mother ship for *Alvin.* Because we had discovered *Titanic* with only four days left in our 1985 expedition, we needed to go back and photograph the wreckage more thoroughly. Most important, I wanted to send a robot inside.

Before going, I did everything I could to patch up my long relationship with the French, which had been destroyed when Steele released the footage to the U.S. press before the French could release it in Paris. I signed over the fee I received from National Geographic for the article I wrote about the discovery and insisted that Jean-Louis be listed as the second author on the article. And I invited him to join us again in 1986. At first the French agreed, but then at the last minute, they pulled out. It felt like I was dealing with a jilted lover who couldn't get past the original heartbreak.

This time, Dr. Steele and Woods Hole were at least publicly supportive. A committee appointed by Steele recommended that he approve the new center to run my explorations, though not without the usual harrumphing about how popularizing science might turn off traditional funding sources. Steele realized that *Titanic* could be a money raiser for Woods Hole, though, and he seemed willing to hold his nose to exploit it. I'd even met with some of Woods Hole's wealthy donors in June 1986, hoping they would support my new center.

The Navy was underwriting this expedition, just as it had the first. We also had a new player on board: *Jason Jr.,* nicknamed JJ—a prototype for *Jason,* a larger version that was still on the drawing board. JJ would be deployed from a submersible, connected by a long, snaking cord so it could go places the larger one dare not try to go. It had evolved from the old Navy robot I was testing off Mexico when I got kicked out of the country, and it was designed to sneak into wrecks and peek at interior spaces that had been impossible to reach.

Once again, I was right in the middle of a top secret operation. I'd be testing JJ on *Titanic,* using it to see if we could get the dramatic glimpses of the Grand Staircase that Secretary Lehman had described when he was pitching my ideas to President Reagan. Reagan saw the military potential of such a robot, too, and he liked how nervous it would make the Soviets when they learned of this new capability. But the Navy also had a second, and more immediate, objective in helping us develop a more nimble vehicle. If *JJ* could penetrate the *Titanic* wreck, then the Navy could use it to get inside *Scorpion*'s torpedo room to learn more about why that sub had sunk. Several officers from Submarine Development Group One, which operated the Navy's most secretive spy subs and its own deep-diving submersibles, planned to go with us on this next *Titanic* expedition to learn how to operate the robot.

So the stakes were high, especially because *JJ* had never been anywhere near the deep ocean. We'd tested it mainly in a swimming pool, and my engineers thought it was nuts to plunge *JJ* into such an assignment without more testing. But off we went. *JJ* would be getting its inaugural run down *Titanic*'s staircase.

Some of my old friends in Woods Hole's *Alvin* group thought I was crazy, too, abandoning the promise of manned submersibles just as others were beginning to use them. Here I'd helped pioneer *Alvin* for deep-sea exploration, and now I was talking up the role of robotic systems for revisiting *Titanic*. *Alvin*'s all-too-human pilots and handlers were upset.

I might have gotten a little ahead of myself when I'd told the *Cape Cod Times*, "Manned submersibles are doomed." My ultimate goal was to tether *JJ* to another robotic vehicle, which could provide a setup with stronger lighting and more staying power. But the technology to enable the two robots to work together in that fashion was not quite there yet, so I had to humble myself and request *Alvin* for this run. After we shoved off, bound for the *Titanic* site, the *Alvin* gang made me eat my words—literally. One of the ship's crew members baked a cake decorated with my words, "Manned Submersibles are Doomed," and I humbly took a bite.

—

WE REACHED THE *TITANIC* SITE on July 12, 1986. We had 12 days to explore, and I didn't want to waste a moment. The first morning was picture-perfect, with calm waters and blue skies. It was time for three of us—pilot Ralph Hollis, copilot Dudley Foster, and me—to climb into *Alvin* for a get-acquainted dive, looking for good landing spots on *Titanic*'s bow.

But almost as soon as we started down, things started breaking. First the sonar went out. Then water began seeping into one of the main batteries. Alarms were going off. I knew we were all right as long as we kept monitoring how much water was penetrating the battery pack. At the same time, our bottom transponders were acting up, which meant that no one aboard our mother ship could tell where we were. We were 12,500 feet down and operating in the dark. We couldn't see any farther than *Alvin*'s lights could shine.

I could see tiny particles, the detritus of marine life, flowing north in the deep current. That meant that the current might have pushed us north of *Titanic*'s bow, so I told Hollis to drive south. *Now we're cooking,* I thought. But just then the battery alarms sounded again. Hollis was ready to call off the dive when the navigator on the mother ship suddenly called out that he could point us toward the sunken bow. We were running parallel to *Titanic,* he said, and as we crept closer in our half-blind way, I peered into the gloom. A large pile of mud emerged slowly. We drove around it, and as we did, we began to see a wall of steel towering over us and disappearing into the dark in every direction. We were staring at *Titanic*'s hull. Never had I felt so small.

But *Alvin*'s technical problems persisted, and we hardly had time to absorb what we were seeing. We dropped the weights holding us down and began our two-and-a-half-hour ascent back to the surface.

After a night of repairs, *Alvin* made the next descent without a hitch. As we got our bearings, we shared another extraordinary moment. Out of the viewport, I could see what appeared to be a sharp edge, almost like a knife. It revealed itself slowly, and I couldn't see past it. As we got closer, I realized, *Oh my God, it's the prow.* We were approaching *Titanic* directly from the front, heading straight into the bow. It looked like a ship was coming out of the dark, straight at us. Dead-on. Made us want to get out of the way. Would it slice us in half? Then I realized it's not coming at us. We were going toward it.

As we got closer, I could see trails of rust running down the side of the bow onto the ocean floor. Created by bacteria that feast on rusty iron, long, reddish spikes hung like icicles from the deck railing. There was no way to describe them—until a new word popped into my head: "rusticles," which years later officially became a word in the *Oxford English Dictionary*.

We crossed over the forward deck and saw the monstrous anchor chains. *Titanic*'s magnificent wooden deck had vanished, replaced by a carpet of thin white hollow tubes, the remains of wood-eating shipworms. We landed in two spots—one near the toppled mast with the crow's nest, the other where the wheelhouse used to be. We got great pictures of the ship's telemotor, although the wheel had been eaten by the wood borers. These landings proved that *Titanic* could support *Alvin*'s weight, at least in spots.

We then made several passes over the bow. We also saw portholes with glass still intact. Imagine the violent, wrenching, breaking of the ship; the long ride to the bottom; and the plunge into clay, buckling the bow in several areas—and still some of the glass was holding on. We never saw the name *"Titanic"* on the ship. Everything was obscured by rust, although later examination of our video revealed what might be part of the "c."

Now I knew exactly where I wanted to land *Alvin*—just beside the opening above the elegant Grand Staircase. Once covered by an ornate glass dome, it led down to first-class public rooms and cabins. We could poise *Alvin* at the opening and send *JJ* down the staircase. What better test for our new little robot?

—

THE NEXT AFTERNOON, we climbed into *Alvin* again, and I asked Dudley to land on top of the telegraph room, just beside the staircase opening. Then Martin Bowen, *JJ*'s pilot, took control, manipulating the little robot sub with a reel like that of a fishing rod. We watched out *Alvin*'s viewport as *JJ* neared the swimming pool–size opening above the staircase. This was *Junior*'s first real dive, so we kept him on a short leash, letting out a little cable at a time.

Slowly, *JJ* descended into the hole, disappearing from sight. Now *JJ*'s eyes became our eyes, as we watched the monitor inside, seeing what *JJ* saw.

The staircase, as it turned out, had become little more than an empty shaft, almost all of its wood and elegant decorations having long vanished or collapsed in a giant heap below. We passed the spot on the wall where once hung a bas-relief clock decorated with winged figures representing Honor and Glory. Martin noticed a room with its entry agape. Like a voyeur, *JJ* wanted a peek. Slowly, Martin maneuvered *JJ* in and, there, amazingly enough, an ornate light fixture still hung from the ceiling, slightly askew. It carried a gift from the deep ocean: a sea pen, a delicate creature attached at a rakish angle like a feather on a woman's hat.

The area below was filled with debris: the remnants of the staircase—whatever hadn't been eaten by the borers—and other wreckage. We had only 20 minutes or so to explore the Grand Staircase. *JJ* made it down to the B deck, and we were eyeing the C deck, but time was up. We wanted to stay longer, but we had to head back to the surface.

That was part of my frustration with exploring in a manned submersible: We had so little time to spend on the bottom. *Alvin*'s batteries had to be recharged each night, and to avoid taxing the pilots, we had a rule that we needed to be back on the surface by dinner time at 5 p.m. *Alvin* ascended slowly, and so that meant we had to leave the sea bottom by 2:30 to 3. Talk about bankers' hours. You've just landed on something as fascinating as the moon, and now it's time to leave. That's why I was advocating for a move from manned to robotic submersibles. Robots don't have to make it back for the early-bird special. We could have 24-hour coverage and not risk our lives every time we went down.

Everyone—my whole crew and the Navy officers on board with us—were pumped up when we arrived back on deck that day. We'd proven *JJ*'s capabilities. We watched the video of its descent into the staircase shaft over and over again. I filed a status report to Woods Hole, then did a series of media interviews. Everyone was excited at the idea of seeing the inside of *Titanic*. Certain questions came up again and again. Had we seen any bodies? Had we retrieved any artifacts? I patiently explained that no human remains could have survived in those conditions. Furthermore, we had vowed—and the Navy had insisted—that we would bring up no artifacts whatsoever. As far as I was concerned, doing that would be tantamount to grave robbing.

On subsequent days, we explored the debris field around the wreck. Artifacts were scattered all over the seafloor—silverware, champagne bottles with their corks in place, boots, luggage. No skeletons, thank goodness. Once, a tiny face emerged from the gloom, its small head half buried. My heart went into my throat, and then I realized I was looking at a doll—a reminder of the young lives lost in the disaster.

We came across a steel safe sitting on the seafloor. I had vowed to retrieve no artifacts. But a safe? No telling what might be in it. We used *Alvin*'s mechanical arm to twist the knob, but the door didn't open. Later, we saw that the bottom had rusted away, and nothing was inside it.

We thought most of the debris must have come from the stern. But was there another section of stern somewhere else, bigger than the piece we'd photographed the year before? We'd been using *Angus* to make reconnaissance runs, and photos revealed a large piece of metal, mangled and flattened, about 2,000 feet from the bow. A chunk was missing from it, and the whole thing looked like it had been pulverized.

I was surprised by its length, about 250 feet. It was clearly the stern with its poop deck—the flat, cantilevered section at the back edge where so many passengers had retreated as the ship slipped into the ocean. It was sitting upright, buried about 45 feet into the seabed.

I wanted to send *JJ* under the poop deck to see if *Titanic*'s propellers and rudder were still there. I wanted to put to rest that claim of Texas oilman Jack Grimm that he had found the propeller on one of his searches. But *JJ* was acting up, so Ralph Hollis decided to pilot *Alvin* down there. Even though he was bending the safety rules, Ralph was confident that he could get *Alvin* in there. *Titanic*'s stern was sloping upward toward open water, which meant we could drop weights and escape quickly if we needed to. It was risky but, frankly, I wanted to see what was there, too.

It was an eerie feeling to be creeping forward, watching the hull get closer and closer. We got so close we could see the rivets in the hull. It was one of the few times I've ever felt claustrophobic. Finally, we were able to see it: the top of the rudder, still firmly attached to the hull. But we saw nothing of the propellers. Maybe we had not gone as far as we needed to see them, but enough was enough. On our way up, we landed on the poop deck and used *Alvin*'s mechanical arm to leave behind a plaque dedicated to Bill Tantum, who had died in 1980, and to all those who had perished there.

I wanted to make sure we completely photographed the bow hull, hoping it would yield clues about damage that occurred as the iceberg scraped along the starboard side. Most of that area had sunk into the ocean's sediment, but we saw no sign of a gaping hole or gash. We did find some buckled and separated steel plates with the rivets sprung. So perhaps, rather than creating a huge gash, the iceberg separated the plates at their joints, popping the rivets. The force from the collision might have resulted in something more like unzipping a zipper rather than bashing in a hole, creating intermittent openings mere inches wide that still allowed water to penetrate a wide swath of the bow.

It left some of us wondering if fewer compartments would have flooded—and if those aboard might have survived—if *Titanic* hadn't veered away from the iceberg but instead had hit it head-on.

—

AFTER WE RETURNED IN LATE JULY, we held more press conferences at Woods Hole and National Geographic. My discovery of *Titanic* took my relationship with National Geographic to a new level. I had written three magazine articles, published a book, and produced two television specials with them. Now National Geographic and Turner Broadcasting, which aired their documentaries, were planning a big show on our *Titanic* expeditions. I also signed a hefty contract with Warner Books to produce an account of the search.

First, though, I needed to clear my head, so I retreated with Margie and our sons to her father's cabin on Whitefish Lake in Montana, where we had been going every summer. After the intensity of an expedition, I'd usually feel exhilarated but depleted, and Montana, with its vast skies and pristine waters, was an ideal place to unwind. There was nothing like firing up the motorboat and streaking across the lake with the boys hanging one-handed on water skis behind me. It was important to reconnect with my family.

Todd was heading into his senior year at Falmouth High, and Dougie was going to be a sophomore. Both had a wild energy—

clearly chips off the ol' block. They were both fanatical about hockey, and at home in Massachusetts, they'd practice in the basement for hours at a time. They had filled hockey pucks with lead to help develop their wrist strength, and Margie and I would hear *twack, twack, twack* as they slapped shots against a goal they'd painted on the concrete wall.

During the school year, I juggled my schedule to make most of their games. The hockey crowd in Falmouth was mostly working class and of Portuguese descent, and Margie felt comfortable in the community. She and I were still kind of coexisting, but we bonded over the boys and their love for hockey. In fact, the first lecture I gave after I found *Titanic* was a benefit for the Falmouth Youth Hockey organization. Dougie said some of the girls were so impressed with what I'd done that he got a few dates out of the deal.

All three of them shared my pride in the *Titanic* expeditions. I recognized and appreciated the sacrifices they'd all made to let me pursue my dream. This Montana vacation was a chance to make up for missing so much time with them, and I gave them my full attention.

I also had some fun as I eased back into work. I was the Navy's cover story as submarine officers who had been on the *Titanic* cruise loaded up *Alvin* and *JJ* and headed back to the *Thresher* and *Scorpion* sites. They wanted to check for any signs of radiation leaks and use *JJ* to look inside *Scorpion*'s torpedo room. My job was to be as far away—and as visible—as possible, in case the Soviets were tracking me after all the publicity about our deep-sea exploits.

As the Navy team transited to the *Scorpion* site in late August, I was in Las Vegas doing early publicity for a video National Geographic was making on the *Titanic* expeditions. Vestron Video, which was going to sell VHS cassettes of the TV show, had flown me to a trade show there, and I found myself in a hotel conference room with Robin Williams and Billy Crystal, who had videos of their own to promote. When I told them who I was, Robin instantly launched into a riff, just for the three of us, running over to the window and barking out orders to an imaginary crew: "Remember: Women and children first!" Billy talked him down, then we all had lunch together.

It would have been easy to get a big head, moving in such company. But just moments after that encounter, we were all on the convention floor near several booths where the provocatively attired stars of adult entertainment videos were autographing photos of themselves. That helped put everything in perspective.

JJ did a good job for the Navy. They found no radiation at either submarine site, but the opening to *Scorpion*'s torpedo room was blocked by a bulkhead, so *JJ* had not been able to go in. Nonetheless, the submarine force was eager to exploit my new technology to gather intelligence that could help deter the Soviets, and I formed a company, Marquest, to develop underwater imaging systems and build an *Argo/Jason* system for Navy spy operations. Rear Adm. Dwaine Griffith, who ran the deep submergence division, was eager to take advantage of the capabilities, and Secretary Lehman actually swore me in as an active-duty commander to help maintain secrecy. I still can't talk about this highly classified work today.

Other new opportunities were coming my way. Here I was, going wide again. Invitations to speak on college campuses were rolling in at $10,000 a pop. That was money that went right into my pocket, the first time I'd earned such serious sums. I hired a Harvard graduate student, Ann Pellegrini, to do some research on one of the new explorations that interested me—ancient trade routes in the Mediterranean Sea—and I began talking to educators about how I could share my kind of science. How could I respond to those thousands of letters I got from kids excited about *Titanic?*

I also found a new passion in genealogy. I'd received a rush of letters from Ballards all over the country who wondered if they were related to me. After giving a lecture in North Carolina, I dug into the archives at Guilford College in Greensboro. It turns out that eight generations of Ballards had been Quakers there—and most of them had signed their names with X's—before my great-grandfather had moved to Wichita.

I was amazed that someone like me, who had invested so much of his life in military service, had come from a family of Quakers. Indeed, although I was able to trace the whole, long Ballard family saga back to the sheriff of Nottingham in 1325, I could find only one ancestor who wore a military uniform, Col. Thomas Ballard, a member of the British colonial militia in Virginia.

I wanted to head off in new directions, but I faced a real problem: how to raise the money. Institutions like Woods Hole provide prestigious places for scientists to hang their hats, but each scientist has to arrange his or her own funding. Traditional sources, like the Navy and the National Science Foundation, provided grants to cover my salary, the salaries of my team members, and the enormous costs of conducting research and expeditions. But neither of them had an interest in the new things I wanted to do, like searching for ancient ships or responding to schoolchildren.

That's why I'd proposed creating the Center for Marine Exploration at Woods Hole. As its director, I could make searching for ancient ships and educating schoolchildren part of my job. Dr. Steele was still dragging his feet about approving the center, and I was eager to get it going so I could use its imprimatur to help me raise money from new sources like wealthy individuals and corporations.

I didn't know many wealthy individuals other than a few I had met through Charlie Hollister, our dean of graduate studies, who had already started cultivating private donors to expand the institute's own resources. I liked Charlie, and I was hoping that by helping him, I could get his help in raising some of the money needed to start up my ventures.

I spoke to a group of business leaders in Newport Beach, California, that Charlie had targeted. He called them the duck hunters because they regularly got together to do just that. They had the connections and the money to make things happen. Charlie would hook them with tales about the most exciting work our scientists were doing, and at each event, he showed a film clip about my discovery of *Titanic*.

In early November, Charlie sent out a memo saying that he had raised more than $400,000 for Woods Hole. He credited my talk in Newport Beach as a major factor in his success. But how much of that $400,000 did the institution set aside for my projects? Not one cent.

In the horse world, there's something called a teaser. When a mare is about to be paired with a prized stallion, another male horse—the teaser—is sent in first. If the mare attacks or kicks the teaser, they don't send in the prized stallion.

That's what I felt like: the teaser. Woods Hole would trot me out to pump up the audience, and then they'd swoop in and control the proceeds. And I got nothing.

I knew I needed to do something to change the situation. I thought of the advice my father once gave me: To have many masters is to have none. In other words, if no single person or organization controlled me, I controlled my own fate. I needed to develop my own set of donors so I wouldn't be dependent on Woods Hole.

But how to kick this next phase into action? The path forward would begin on a trip to Mexico.

CHAPTER 8

MY CINDERELLA STORY

I had never been on a duck hunt. When Don Koll, the California real estate developer and Woods Hole donor, invited me to join one in Mexico, I wasn't sure what to wear. So I fell back on my Army training, went to a surplus store, and bought a camouflage uniform, military boots, and a camouflage stick to paint my face.

We gathered outside the hotel in Culiacán early one morning in January 1987. It didn't take me long to realize that I had gone too far. It was so dark we could barely tell each other apart, and when Marco Vitulli, Don's business partner, waved his flashlight around to identify everyone, he broke out laughing as the beam crossed my face.

"My God, Ballard, you look like Ollie North," he said. North was the Marine officer who had just become notorious for his role in President Reagan's Iran-Contra arms deal. The moniker stuck. For the rest of my many years with Don, Marco, and the other duck hunters, I was no longer Bob. I was simply "Ollie."

After driving nearly an hour to reach the airboats, we wound our way through a maze of small channels. The pink sky was teeming with what seemed like hundreds of thousands of birds. When we finally arrived at the wooden blind hidden in the tall reeds, Marco loaded his Benelli 12-gauge shotgun, adjusted his sunglasses, and scanned the horizon.

I'd been an expert rifleman in the Army, but I'd never used a shotgun before, and I couldn't see how to load it. The gun that the club had provided me was much older and different from Marco's side-loading model.

"I think you load it from the bottom," Marco said.

By then, it was almost completely quiet. The birds that had swept across the sky on our journey to the blind were gone, scared off by the roar of our airboat. I braced my gun barrel up between my legs and fumbled to load the shells from the bottom.

Without warning, the gun went off, straight up into the sky.

I froze, knowing all too well that I had broken the cardinal rule of hunting and forgotten to keep my safety on.

Then we heard the distinct flap of wings and saw a duck lying on its back in the marsh in front of us. It was a male cinnamon teal, considered the superfast fighter pilot of ducks because it races low over the marsh and is one of the most difficult ducks to shoot.

I looked over to our bird boy—the guy who'd set up the blind and would retrieve our ducks—and he was valiantly trying not to laugh. Marco just grinned and shook his head at me.

"No one is going to believe this," he said.

The next morning, I found myself in another duck blind with Don. "So," he said as we neared the end of our shoot, "what do you want to do next, now that you've found the *Titanic*?"

I had my answer ready: I wanted to go much farther back in time and search for ancient shipwrecks in the Mediterranean and the Black Seas.

"How much would that cost?" he asked.

Normally, going to sea with my kind of technology requires a large ship and about a month to conduct a reasonable hunt, I explained. "We're talking about one million."

Don tilted his head, thought about it for a split second, then gave a decisive nod.

"I'll give you a million, as long as you let me tag along," he said.

"Done!" I exclaimed.

But my vision didn't end there.

I've always been a two-or-three-or-four-birds-with-one-stone kind of guy, and I had a plan to make the most of this venture, even if it might take two to three years to pull off.

Our National Geographic television special, *Secrets of the Titanic,* had received one of the highest ratings ever for a cable television documentary. National Geographic and Turner Broadcasting were eager for another round. My plan was to charter a ship in England and head into the Mediterranean to search for ancient vessels. That meant we would be sailing down the west coast of France near where, after an intense battle with British naval forces, the German battleship *Bismarck* had sunk at the beginning of World War II. More than 2,000 German sailors went down with the ship. Though it wasn't an ancient ship, diving on *Bismarck* would make great television. I was delighted when they agreed.

I'd also been thinking about those thousands of letters I'd received from schoolchildren wanting to be like me after I'd found *Titanic.* I felt a moral obligation to respond in a big way, remembering how a kind dean at Scripps answered my letter in high school and started my career. Why not use my robotic cameras—that idea of telepresence—to take kids along on the expeditions? The National Science Foundation had not shown much interest in funding the technology needed to beam live shots from the robots back to people on land. Maybe I could use the idea of educating children to finally make this a reality.

It was clear that a crisis was developing in science education in the United States. Even in the 1980s, fewer than half of all science and technology graduate students in U.S. colleges were American, and the country I loved had a huge deficit of scientific literacy. I wanted to counter the peer pressure that causes middle school children, especially girls, to lose interest in science. I had no doubt that kids who were obsessed with *Star Wars,* R2-D2, and video games would be excited by my robots and undersea expeditions. If I could transform their fascination with the *Titanic* discovery into a greater interest in science and engineering, that would really be something. I even had a name in mind. I wanted to call it the JASON Project, in keeping with the family of robots I was developing.

I knew I would probably have to tack with the wind a few times to realize my vision, just as the mythical Jason had done with his *Argo*. Nothing comes easily when you're thinking outside the box. Even if my bow wasn't always pointed at the finish line, I felt like I could still get there, and maybe my greatest legacy would be creating a new generation of scientists and explorers.

—

I CLEARED SOME OF MY SCHEDULE that summer for time with my sons. First, I took Todd and Dougie rafting down the Colorado River. Dougie had just finished his sophomore year at Falmouth High. Todd had just graduated, and I could tell he was struggling to navigate the pressure coming his way from all the *Titanic* attention, especially because he wasn't interested in following in my footsteps.

Todd had blossomed into a strong hockey player. He loved cars and was fixing up an old Mustang in our barn. A fearless teenager, he lived for speed and risk. Still, he was a gentle kid who didn't take to violence and did everything he could to avoid a fight. Even though he was almost as tall as me and sturdier, his little brother thought of him as a bit of a wimp. But when Todd stood up to a bully of a football player who was threatening one of his smaller hockey teammates, Dougie's opinion of his older brother changed.

After being pushed back, literally to the edge of a cliff, Todd had fought back, knocking the football player out cold with a single punch. He called me and said he thought he'd killed the guy, and when I raced over, the place was lit up like Coney Island with the lights from ambulances and police cars. Luckily, the football player came to. Later, when I heard the guy wanted a rematch, I gave Todd some sound fatherly advice: Pass the word around that if you fight again, he needs to swing first, because if you accidentally kill him with your counterpunch, you don't want to be held responsible for his death.

Needless to say, the football player backed off.

I was proud of Todd, if a little worried that he was going to push his love of adrenaline too far. A month later, Marco Vitulli took Todd and me on a salmon fishing trip in Canada, and I hoped some of Marco's confidence and business focus would rub off on my rambunctious son. I should have known better. Marco was the kind of guy who tells jokes perfectly, the life of the party. He and Don Koll had gone to Stanford together and then into the Air Force, flying jets in the Korean War. Instead of calming him, Marco let my daredevil son fly his De Havilland Beaver seaplane—which, of course, Todd loved.

After our vacations, I began to assemble the pieces and the teams I would need to test my concepts and achieve my new dreams. I have to say, I felt a little tentative, like Cinderella getting ready to go to the ball to meet Prince Charming. I needed a dress, carriage, coachman, and attendants to get me there—and a lot more time than just to the stroke of midnight.

Marco was going to join with Don in putting up the million dollars of seed money, and we were trying to figure out how to structure it as an investment rather than a gift. Dr. Steele, the Woods Hole director, had finally approved the creation of my Center for Marine Exploration, but he allotted only $20,000 to cover the most basic expenses. I needed to get creative to finance my new expeditions, and I was willing to share film rights and royalties with Don and Marco so they would get a return on their investments.

I needed a satellite link from ship to shore, and I needed technical help to produce live broadcasts for students. I needed educators to develop a curriculum of basic science that students learned before our shows, and I needed museums and science centers that could provide auditoriums where the students could watch. I needed to find an ancient ship to explore in the first educational offering—not to mention that I needed to find *Bismarck* for a paycheck to make this all happen. I was working on all these tracks at once, trusting they would converge in the end. My book on *Titanic* also was coming out, and I had to take off the month of October to promote it on a tour across North America and Europe.

I was giving a lecture at Southern Methodist University in Dallas that November when I got a break. I mentioned my dream of using satellite technology to link my ship's explorations live to kids in museums around America. As fate would have it, one person listening was Jim Young, who worked for EDS—the electronic data company founded by Texas billionaire H. Ross Perot. After the lecture, Jim and his wife, Carole, came up to me. In his wonderful Texas accent, he said, "I think we can help y'all."

Jim and a colleague, Diane Spradlin, came to visit me at Woods Hole in the dead of winter—proof of their serious interest in supplying the satellite link I needed. Then Lester Alberthal, Jr., the new president and chairman of EDS, followed up with his own visit, saying he just wanted to meet me. "Don't worry—we're in!"

With that encouraging news, I felt confident to set off to the Mediterranean in search of ancient shipwrecks. M/V *Starella,* a research vessel, departed from England in May 1988. The preliminary work that Ann Pellegrini had done for me made me think it wouldn't be hard for *Argo* to find an interesting site. Then we would go back the next year prepared to film and broadcast the expedition for school kids. Because this was just a scouting expedition, I decided to keep costs down by hiring a younger and less experienced team than I normally would have.

I was always looking for ways to add a bit of paying work to any expedition, so I agreed to a quick side trip as we entered the Mediterranean, to help Lloyd's of London, the big insurance company. A client of Lloyd's claimed his ship had struck a submarine near Spain and had sunk. Lloyd's questioned the claim, because all the crew members had fully packed luggage when they boarded the lifeboats. They wanted me to collect any evidence I could about the incident.

We were passing by that very spot, and I figured the assignment would give my crew a chance to practice. As luck would have it, we lowered *Argo* directly over the bridge of the sunken ship and towed our camera sled back and forth for 12 hours, scanning every surface we could, taking black-and-white video and color photographs. Every image revealed a nice, smooth hull without a single scratch, no damage in sight.

I was told later that the captain was quite surprised to be shown images in court of his perfectly preserved ship sitting upright on the bottom. Clearly, the ship had been scuttled to collect insurance, and our underwater imagery proved the point. Given our role in exposing the fraud, I only hoped that my name wasn't mentioned.

—

I BROUGHT TODD ALONG on that Mediterranean scouting expedition, his first cruise with me. He had just finished a post–high school year at Brewster Academy in New Hampshire and was preparing to head off to Northeastern University. He still exhibited those wayward signs of youthful abandon that I found both charming and concerning. I hoped that the disciplined routine of life at sea would temper some of his impulsiveness.

I had never worked in the Mediterranean before. We started by sweeping the bottom contours of the Tyrrhenian Sea, off the west coast of Italy. Being there felt a little more intimidating than it had when I had just studied the giant map hanging on my wall. As the days passed by, our searches in several likely areas found no signs of any artifacts, and some of the younger and less experienced crew members started murmuring about whether I even knew what I was doing.

We did spot numerous amphoras—the large ancient clay jars used as containers for wine, olive oil, and other goods. They littered the seabed on Skerki Bank, west of Sicily and north of Tunisia. There were so many that we said we'd found "amphora alley," an ocean passage to Rome. Then we spotted the shadowy outlines of my first ancient shipwreck. She sat in 2,700 feet of water. Much of the ship had been eaten by wood-boring organisms, but thanks to *Argo*'s cameras, we could make out the shape of the hull through a series of depressions filled with artifacts—the large amphoras and pottery that had been her cargo. We later figured out that she dated back to the third or fourth century A.D. We named her *Isis*, after the Egyptian goddess who protected seafarers and helped the dead enter the afterlife.

I was thrilled, but the novice crew members were expecting to find something more substantial. Some were visibly disappointed. They seemed more interested in having a good time than in knuckling down to work at the level of detail needed to monitor our searches. For the first and only time in my long career at sea, I called off the Mediterranean expedition early. My only consolations were that I had located the ancient ocean highway and a wreck along it—and that I would be returning with my A-team in the following year to do more exploring.

—

THE PLAN WAS TO SKIRT OUT into the Atlantic to hunt for *Bismarck* the next month, as *Starella* steamed back toward England. I hired a more experienced crew, hoping that would reenergize me—and inspire Todd. With my reputation—or at least a bit of my dignity—hanging in the balance, I grew worried as we faced rough seas and poor weather on the way there.

Bismarck was one of Hitler's mightiest battleships—and like *Titanic,* thought to be unsinkable. In May 1941, it embarked on its first mission, Operation Rheinübung, with the goal of intercepting British shipping lanes in the Atlantic and sinking as many boats as could be found. After suffering a series of strikes from air and sea, *Bismarck* sank, presumably a few hundred miles west of Brest, France. So that's where we were headed. The ship probably lay in deeper water than *Titanic,* and the search box was larger, but I had boasted we would find her quickly. National Geographic producers had come along, expecting to film a triumphant expedition. We would come back the following year to the site and explore it again, explaining what we were seeing and sending live video to kids watching in schools and museums across the country.

It wasn't that easy. Tempestuous conditions persisted throughout the voyage, and the undersea terrain was tricky. The search area included a hilly bottom and an underwater mountain that made piloting *Argo* precarious. I was nervous about smashing it into a cliff and losing *Argo,* so we just searched the flattest areas.

We spent four days tracing ghost trails of debris that looked like they might belong to *Bismarck*. We saw an impact crater that could have been created by a heavy part of a ship that had fallen in the midst of battle and slid down, causing an avalanche. I felt a flicker of hope, but cautioned myself. Todd felt certain we had found her. I wanted this victory for him.

We swept the crater for hours. I ran all the test calculations, reread the history, and pored over the details. But *Bismarck* remained elusive. The hours ticked by and the watches felt like an eternity. Mechanical problems pushed our equipment to the limit. Making matters worse, Todd was challenging my authority. Maybe he was just showing off for the two friends who had come along with him, but he seemed sullen, and he showed up for his watches late.

I felt defeated. I ordered *Starella* back to port, saying we'd take a closer look at the color photographs as we headed home. But once they were processed, it was like a stake in the heart. There, instead of any piece of *Bismarck,* was a beautifully preserved teak rudder, obviously from a 19th-century sailing ship, clearly not part of a battleship at all. It felt like the perfect failure, both personally and professionally.

It was the first time I had ever returned from a mission without anything to show for it. I could usually brush off any setbacks, but missing *Bismarck* was really humbling. I also began to fear that it might jeopardize my fundraising. When our National Geographic producer, Chris Weber, asked how I felt, I put the best face on it: "Round One to the *Bismarck*. I know where it isn't. I'll get it next time."

—

THANKS TO NAVY FUNDING, we spent the rest of the year developing *Jason* and *Hugo*—bigger, better vehicles to take back to the

Isis site in 1989. Weighing over a ton, *Jason* would be my most sophisticated ROV (remotely operated vehicle) yet, with three video cameras, a still camera, its own sonar, and two mechanical arms and claws that could pick up artifacts. *Hugo,* named for "huge *Argo,*" also served as a giant garage for *Jason,* designed so we only had to launch one object. We would lower *Hugo* to the bottom and then deploy *Jason* from inside it. When the time came to bring them back up, we drove *Jason* inside *Hugo* for the return to surface.

We also were testing a fiber-optic cable that could carry color video up to the mother ship as *Jason* took it. For the last four years, *Argo*'s black-and-white video had guided our explorations, but we needed color if we were going to capture children's imaginations.

I was counting on National Geographic to help me out. Tim Kelly, head of National Geographic Television, agreed to fund a return expedition to *Bismarck,* but he said his team was not equipped to do live broadcasting, and I would have to talk to Turner Broadcasting about that. So I went down to Atlanta and met Ted Turner, the prickly media entrepreneur who had started CNN and built TBS into a national superstation. Luckily, Turner is just as filled with energy as I am, and he loves the sea.

When I described my idea of broadcasting my underwater explorations live to kids around the country, Ted got it. "You need to go down the hall and meet my team that does the Saturday morning wrestling shows," he said. *You've got to be kidding me,* I thought, but off I went. When I told his wrestling producers what I had in mind, they were so excited, tears came to their eyes. Now I needed to build the audience.

Representatives from both the National Science Teaching Association and the National Council for the Social Studies had agreed to create the science courses for the kids coming to our shows.

Now I needed to find museums to host our broadcasts over a two-week period. I was imagining as many as 250,000 students.

Organizing something audacious like this seems to be my calling. The first part is to have a crazy idea—like the ones I'm constantly dreaming up. Next, you assemble a group of organizations that are all impressed by the others sitting around the table. Usually all you need is for one of them to say, "I'm in," and everything else falls into place.

So off I went to the Boston Museum of Science to meet the president, Roger L. Nichols. The museum had given me its prestigious Washburn Medal in 1986, so the door was open. It took about two nanoseconds for Roger to sign up, and another few seconds for him to get on the phone to his network of museum presidents. Before I knew it, a dozen downlink sites across the United States and Canada had signed on—and this was even before the internet!

And, after all our clashes, John Steele agreed that Woods Hole would be one of the sponsors of this first *Jason* expedition. In November, we held a press conference at Woods Hole announcing the initiative and our plans for the first live broadcasts ever from the ocean floor. We promised 84 broadcasts—six a day—from the Mediterranean in May, just six months away. And to think this was all still just a giant experiment. All that was left was for Cinderella to go to the ball.

—

WE SET SAIL FROM GIBRALTAR on April 22, 1989. I made sure I had a better ship this time—*Star Hercules,* an offshore supply ship—and an all-star crew, including Andy Bowen, Dana Yoerger, and Cathy Offinger, all Woods Hole colleagues and veterans of *Titanic* expeditions. Cathy was my logistics chief and overall right hand. I even created a business card that said "Robert D. Ballard—

Explorer—Call Cathy." Our plan was first to broadcast from the Marsili Seamount in the Tyrrhenian Sea, to show an active underwater volcano, and then move to the *Isis* site, 110 nautical miles to the south. Before we went live, we needed to conduct a mock broadcast with the team at TBS in Atlanta—a dry run and low-stakes rehearsal to be prepared for our live takes a few days later. The weather was getting worse, though, and I was worried it might jeopardize our equipment. With my mic on and the cameras rolling, I kept sneaking peeks at our shipboard monitor at the same time that I welcomed our supposed viewers to watch us launch *Jason* and *Hugo*.

I had just finished saying that the cable lowering *Hugo* was unspooling nicely when a large swell came through. The ship heaved up on the wave and then suddenly dropped down into the passing trough. When the ship went up, so did *Hugo,* with *Jason* inside. But when the ship went down, *Hugo* and *Jason* didn't follow as fast. So when we rose up on the next swell, *Hugo* and *Jason* were still sinking—and the cable connecting them to the ship snapped.

"Oh, my God!" I shouted. I could just see it. Millions of dollars of equipment in a free-fall descent, nearly 3,000 feet, to the bottom of the sea. I ripped off my headset and ran out onto the deck. Everyone was just standing there, looking over the side of the boat, stunned, as if they could see the equipment plummeting.

My military nature kicked in. I started giving orders. I shouted to my top electrical engineer, Bill Hershey, to haul out the little vehicle we had used to test our new fiber-optic cable. We could add lights, a camera, and a grapple, and maybe it could help us pull *Jason* and *Hugo* off the bottom. In the midst of all this, Dana asked if I wanted to know how fast *Hugo* and *Jason* were sinking. I couldn't believe that was where his mind went. We needed action, not analysis. The ROVs were going to the bottom, no matter how fast.

Bill and his team of grease-splattered engineers sprang into action. We hastily dubbed our backup *Medea,* after Jason's wife, even if she did kill her own children in the myth. The crew got her ready to go within 24 hours. If we lost *Medea,* we had nothing. The transponders on *Jason* and *Hugo* were still sending signals, thank goodness, so we knew exactly where they were. It was risky. The weather was improving but remained perilous. Two days before show time, April 29, it was all or nothing. I gave the order, and down *Medea* went.

We all held our breath and crowded around the monitor, anxiously watching the images coming from *Medea.* Dana did a great job navigating the ship as we came down right on target. Fortunately, *Hugo* and *Jason* were sitting upright, with a chain bridle visible on the top of *Hugo.* The plan was for Martin to lower the grappling hook suspended on a chain from *Medea* and slowly inch it over to *Hugo*'s bridle. Just at the critical moment, another large swell came through, lifting the grappling hook off the bridle. Then, somehow, on the next swell, the hook grabbed onto the side of *Hugo* and would not let it go. It wasn't perfect, but it would have to do. I gave the order to come up as fast as possible, and with one big yank, *Hugo* came off the bottom in a cloud of mud.

When all three vehicles reached the surface, it was sheer madness for about 20 minutes. We secured *Hugo* to our stern, but *Jason* slid out of its garage and floated under us, heading for the ship's propeller. I didn't have time to tell people what to do next, so I ordered the crew to launch the rubber Zodiac. *We're not going to lose this one,* I declared to myself as I jumped into the Zodiac, yelling for someone to throw me a recovery line. We finally got a secure line attached to *Jason* and reeled it in.

After that, the live broadcasts seemed fun—a relaxing break from the anxiety of almost losing all our equipment. I sat in the control van with my headset on, talking the students through what they were seeing, answering questions, and explaining telepresence. *Jason*'s cameras showed students the water shimmering out of a hydrothermal vent near the volcano, and then we moved on to the *Isis* site, giving students across North America a glimpse of the shipwreck and how we mapped it.

I had invited Anna McCann, a pioneer in underwater archaeology, to sail with us to assess the *Isis* site. As our cameras scanned the artifacts, she advised on which ones to recover. Students got to watch as we collected 48 objects from *Isis* and 17 amphoras from other spots. *Jason*'s claws picked them up gently and deposited them in a mesh-bottomed elevator assembly that we hoisted to the surface.

The JASON Project was working. Our broadcasts of the *Isis* expedition were a huge success. Now I had a way to share the excitement of marine archaeology with hundreds of thousands of kids.

Meanwhile, Don and Marco were willing to keep funding me for a second shot at *Bismarck,* and I had money from the Navy to test *Argo* at much greater depths than before. National Geographic also was willing to hold off on the show and keep the cash flowing. Feeling encouraged, I was ready to head back to the Atlantic and take on *Bismarck* for Round Two.

—

AN OLDER, SEEMINGLY WISER TODD joined the crew again, along with two of his friends from Montana, Billy Yunck and Kirk Gustafson. It made me proud to watch their skilled hands and sharp eyes as they took turns piloting *Argo* through the underwater darkness.

After 10 days of looking, we were still seeing mostly mud. I was trying to unwind, playing a boisterous game of Trivial Pursuit, when someone called my name, saying they'd found some debris I should look at.

I vaulted out of my chair and booked it to the control room. There on the video screens was a cluster of small black objects that seemed man-made. Were they part of *Bismarck*'s debris field? As we tracked them, the nature of the bottom suddenly changed from flat and smooth to highly disturbed, with rocks and sediment all over, as if strewn by some great force. I tried to visualize what must have happened here. An underwater landslide. Only something as large as *Bismarck* could have caused an avalanche like that.

We kept combing over this area, searching for anything of interest. I kept my eyes glued to the monitor, trusting my instincts. Two days later, after 48 hours of relative sleeplessness, I saw the telltale shape of a gun barrel coming out of the gloom—pointed right at *Argo*. We jerked the camera sled up to avoid a head-on collision. But there it was. We had found *Bismarck*.

Todd rushed to my side at the monitors as soon as the cry of discovery rang out. His face was burning with pride and excitement. Father and son together, we were going to explore a wreck that hadn't been seen for half a century.

As *Argo* crept over the ship, we saw the holes in the deck made by British shells. But the hull still looked intact—no damage, no holes, no clear reason for its sinking. A few German survivors had claimed that the commander had chosen to scuttle the ship, sacrificing it rather than letting battle damage bring it down. What we saw seemed to validate those claims.

Then, through the haze, we saw the mark of a giant cross slowly come into focus, splashed across the stern.

As we grew closer, I realized it wasn't a cross at all. It was a swastika: That haunting symbol that signified devastation and hatred for so many tempered our excitement. We were peeling back the layers of a dark and ugly history, and we did our best to tread lightly.

Our return to shore was bittersweet, with the horror of the past we had just witnessed still raw and fresh in our minds. Still, we were a team triumphant when we got back home. Our discovery launched another media blitz. Todd came with me to the press conference at National Geographic. He was on top of his game—mature, confident, professional. I had never been prouder.

—

JUBILANT AND VICTORIOUS, with their new explorer bona fides established, Todd, Billy, and Kirk joined up again in Montana, water-skiing and enjoying the freedom of no longer being cooped up on a ship. A Geographic film crew also came out, working on a show about how I had pulled the JASON Project together.

At one point, an interviewer asked Todd what it was like to be the son of Bob Ballard.

"He's a different father," Todd said. "He's more laid-back. He lets me do what I want, and I like that freedom. It feels good."

Then he added, "Out at sea on the *Bismarck,* it was great . . . I was part of the action. I wasn't treated like Dr. Ballard's son. I was treated as part of the team, and it was really fun."

CHAPTER 9

LAID LOW

One night in late July, Margie and I woke to a pounding on our cabin door in Montana. It was two in the morning, and I was dead asleep. I stumbled out of bed and blearily answered the knocking. A sheriff's deputy from nearby Whitefish stood at the door.

In the dark, his young face looked drawn, his expression strained.

"Dr. Ballard," he said. "I'm sorry to tell you that your son is dead. And so is his friend."

He must be mistaken, I thought. Dougie had gone out with friends earlier that night, but by now he was asleep in his room, I assured the deputy.

But the deputy's face grew more pained. "It's not Douglas," he said. "It's Todd."

In that instant, I could feel my world collapsing.

His words brought me to my knees. I found myself winded, clutching Margie's hand. The deputy told us that Todd and his friend had died instantly that night in a car crash on a back road in Cape Cod.

Fresh from the *Bismarck* expedition and the media appearances in May, Todd and I had joined Margie and Doug for the rest of the summer together. I had taken off for a quick business trip, and when I returned to Montana, Todd was gone. Margie told me he had headed home to Massachusetts, on a mission to fix the fraying relationship with his girlfriend, who had grown impatient with his absence and broken things off. Todd, ever impulsive, had driven his old Ford LTD across the country to try to make things right.

Dougie, Margie, and I flew back to Massachusetts right away. We slowly pieced together the last few hours of our Todd's life.

Todd had spent some time with his girlfriend, mending their relationship. Now in good spirits, he and Chad Dalton, his old roommate from Brewster Academy, had grabbed a few beers and were driving home too fast, heedless of the recent rainfall and poor visibility. When Todd came to a slick stretch of pavement in a curve, he lost control. The car slammed sideways into a tree and flipped. Both boys were killed on impact, their necks snapped.

One more piece of the puzzle: When we pulled into our driveway in Hatchville, I realized that the new Ford Thunderbird I had bought Margie—the car Todd coveted, the car we had forbidden him to drive—was gone. He had died in Margie's car, not the hand-me-down family car he had driven across the country. He must have found the keys and decided to take her for a spin that fateful night. Todd didn't know the car, and it proved to be more than he could handle. He must have gunned the engine into that slick curve, and it was too much.

When I went to the site of the accident, just a couple miles from our house, the wreckage was gone. But as I knelt to touch the ground where his body had lain, I found the Thunderbird emblem from the front of the car lying in the grass next to the shattered remains of the oak tree. It sent chills down my spine to see the skid marks in the dirt, the final traces of Todd's rich and vibrant life. Todd was wild—always had been. He was fearless, and he loved speed.

It was speed that killed him in the end.

—

I WAS THE ONE to claim Todd's body at the morgue. I went alone. That's what fathers do. They brought him out and unzipped the body bag. It destroyed me. He had been just a few weeks shy of turning 21, on the cusp of manhood. Rigor mortis had set in, and his body was hard as a rock. The only thing still soft were his eyelashes.

The night Todd died, I was driving alone in Montana, and out of nowhere a lightning bolt hit the ground off the side of the highway. It took me by surprise. The night was calm, no other lightning in the sky. I later learned it was the same time Todd had died. Days later, when Todd was being cremated, a lightning bolt struck the crematorium, shutting it down.

I may be a scientist, but there are some things I can't explain.

We held the funeral service at the Lutheran church in Falmouth. Hagen Schempf, a Woods Hole graduate student who had been with us on the *Bismarck* expedition, gave the eulogy. He had watched Todd grow up and had worked alongside him on two cruises, so he was able to reflect on Todd's coming-of-age. Hearing my son's brief life recounted for all to hear was difficult enough. But one part of Hagen's eulogy struck home hard.

Hagen said he had watched Todd step over the line that separates men from boys. "All boys pass that line sometime and only become fully aware of it in retrospect," he continued. "Todd's fight to get there must have been tougher than most. Measuring up to peers, parents, and siblings is no easy feat. But he had jumped out from behind these shadows and was beginning to cast his own. He was able to fight his way to the surface and chart a new course of his own, while retaining all those attributes that endeared him to us all."

We flew back to Montana and had a gathering at the lake. My parents, my brother and sister, and some of our friends were there. Everyone was sweet about it—wonderful, really. A lot of it is a blur. You want to block out this kind of memory.

When you lose a child, you never recover. You don't go through the mourning period and then pick up your life again, as if nothing happened. You arrive at an entirely different place, searching for a way out. Everyone handles it differently, comes to it in one's own way.

Back in Massachusetts, as I was making arrangements for the funeral, a couple had shown up on our doorstep. We'd never seen them before. They said they had lost their son too. We sat together in that house where Todd had grown up, and we four talked. They had been through it all—the stages of grief, the unraveling. It's like a club, only you don't want to be a member. To this day, I don't know who they were. I call them angels. Taking after their example, I visit other couples myself now when I learn someone has lost a child.

Others were less kind. It was bad enough that I lived so close to the spot where Todd had died. I could never push it to the back of my mind. But then I got a bill in the mail from the owner of the tree that Todd hit, demanding I pay for the destruction of his property. I went ahead and paid it. I was too worn out to fight.

When I lost my son, everything began to crumble. Many marriages do not survive the loss of a child, and mine was no exception. I had known for a long time that ours wasn't a happy marriage, and after Todd's death, that was pressing in on me, too. I had realized years earlier that both my brother and I had married our mother—a wholesome housewife whose role in life was to support her husband. But after I began to work with professional women who had their own sets of goals and accomplishments, I realized a true partnership might look different. I loved Margie, but we were never on the same page, and it had been clear for a long time we never would be. I was comfortable in an academic setting, and Margie never seemed to be.

I had watched what Richard's divorce had done to his children a decade earlier and had sworn to stick out my own marriage, at least until my sons left home. I had poured myself into my profession and my kids, resolving to neglect my personal dissatisfaction for a nebulous greater good. I fell into the rhythm of a hockey dad, taking the boys to games and trying not to think too hard about the obvious.

But when Todd died, my world came crashing down, and I needed to escape. Todd's death put it all in bold letters. Less than three weeks later, I told Margie I could not pick up the pieces. I know it seems heartless to have suggested a divorce so soon after Todd's death. But there was no normal to go back to. I was so deeply unhappy.

The separation broke Margie's heart. I remember her holding a picture of everybody and just saying, "I've lost my two men." I'll never forget that. But it was a long time coming, and I wanted out. Dougie was leaving to attend boarding school. I was supposed to be in New York in mid-August to tape my first *National Geographic Explorer* shows with Tim Kelly, head of National Geographic Television. I'm sure Tim could have found someone else, but I told myself I needed to do it to maintain some sense of stability.

Instead, it ended up being my own private hell. We were taping a month's worth of National Geographic shows, both for *Explorer* on TBS and for another series called *On Assignment,* an hour-long show featuring four or five segments on history, science, and technology. My job was to introduce each one. I did hundreds of takes reading off a teleprompter, segments to open and close the show and to take us in and out of commercial breaks. Between takes, I'd go into the back room and cry, and the makeup people would have to fix my face so I could go out there and do it all over again. It was the hardest thing I ever did in my life.

Promotional photos from that time show me trying my damnedest to smile, but my eyes look hollow and dead. I had always shouldered a lot and thrived on constant stimulation, but the frenetic lifestyle that once drove me forward suddenly felt impossible to maintain. How do you crawl out from under? I just buried myself in work until I emerged on the other side.

In the meantime, I did everything I could for Margie as we went through the divorce, from filling out paperwork to hiring a lawyer for her. Much of what I made on *Titanic* went to her, and she kept the Montana cabin I had just built and two lakeside lots nearby, all fully paid for. In many ways, I had known deep down that one day Montana would be where Margie would go when we separated, because she had roots there.

I was questioning everything—calling on God, the universe: Why did you take my son? I couldn't process it. As it happened, at that time Bill Moyers was airing his series of interviews with Joseph Campbell about his book *The Power of Myth*. I started watching the shows, and I was hooked. Then I got Campbell's tapes—conversations full of stories and philosophical gems—and listened to them as I drove. Sometimes I would get to Woods Hole before I was done with a tape, and I would sit in the parking lot just to finish listening. Often, someone would come over and tap on my window to see if I was OK.

Campbell's words gave me something to hold onto when I felt like I was drowning. He talked about how life is a journey, a constant act of becoming. You never arrive. At a time when I was so focused on survival and not losing what little faith I had left, Campbell challenged me. He talked about how we all must proceed on the hero's journey, going forth, being tested in mind and heart, overcoming those tests and attaining truth, and then returning to society to share that truth. That then releases you to go forth on your next journey. We each have to descend into the darkness before we are ready to emerge into the light, to resurrect. We have to confront the darkness somehow. And so I did.

—

I WAS STILL READING parts of Campbell's book when I went to sea that fall. The Navy had commissioned a new version of our *Argo* camera system, and my company, Marquest, was testing it in very deep waters between Woods Hole and Bermuda. I was glad to help check out the system, but more than that, I was relieved for the chance to disappear for a while.

In so many ways, I wanted to start over. But I wasn't ready. I needed time to mourn a lot of things. I had overcome many set-backs in my life, but losing a son and ending a 24-year marriage sends shock waves through you that make you question your very existence and purpose in life. I needed to see if I still had a fire left in my belly.

After I got back from the Navy cruise, I took time off to work in my garden and gather my thoughts. I decided to make my dream of expanding the JASON Project a reality. It was all about the future, motivating the next generation of young minds to fight the fight and overcome setbacks, to get up after being knocked down.

The JASON Project became my life vest, saving me from drowning. Mike Gustafson, a businessman and friend who owned the cabin next to ours in Montana, offered to help me create the JASON Foundation for Education, and Paul Torgerson, his lawyer in Minneapolis, did the legal work. We were going to do our level best to bring the thrill of undersea exploration to school kids everywhere, showing them the pure thrill of science.

I invited both of them to join me on the foundation's board, along with Craig Dorman, my new boss at Woods Hole; Guy Nichols, chairman of the Woods Hole board; Les Alberthal, chairman and president of EDS, who had replaced Ross Perot; and Gil Grosvenor, chairman and president of the National Geographic Society—and son of Melville Bell Grosvenor, who had launched my career with the Society. We also began planning our next expedition, which gave me a renewed purpose.

I remember stopping at an airport conference room a few months later to meet the National Geographic team that had been assigned to help me get the JASON Foundation up and running. I zipped in, gave a presentation, answered some questions, and rushed out. One of the people there was Barbara Earle, the Society's director of special projects for their television division. She was developing new areas of television media for the Society, and Tim Kelly had made her Geographic's point person for the JASON Project. I heard later that her reaction to me was: *Really? Who is this guy?*

Barbara came on a scouting trip to Lake Ontario to prepare for Jason II, the next expedition we planned on filming and sharing with kids through the JASON Project. The plan was to explore the wrecks of two American schooners from the War of 1812, *Hamilton* and *Scourge,* which both sank during a storm in the Canadian waters of Lake Ontario as they prepared to fight the British fleet. Fifty-three American sailors had died. The wrecks had been located by sonar in 1975 and declared a national historic site, managed by the city of Hamilton. We planned to send *Jason* down to take photographs. I was happy Barbara was with us. Her affiliation with National Geographic carried so much weight in dealing with the political side of the puzzle.

We all had dinner together each evening. I found myself able to open up to Barbara about Todd in private, working through some of my grief with a willing listener. I could tell that she was not starstruck by my discovery of *Titanic,* which actually helped a lot. She had worked with plenty of other explorers, and she had an impressive list of accolades in her own right as a television producer, involved in everything from children's programming to political ad campaigns. Her work had taken her all over the United States and to foreign countries, and she had lived in New York before settling in Washington, D.C. She had a suitcase packed 24/7, ready to hop on a plane on a moment's notice. Like me, she lived life at a sprint, and I appreciated the companionship, even as I was laser-focused on getting my work done and pushing past my pain.

Several months later, in March 1990, I set off with Barbara and other members of the *Jason* team on another scouting trip. We were still over a month away from broadcasting the Jason II shows from Lake Ontario, but we were already exploring possibilities for Jason III in Ecuador's Galápagos Islands, where we could feature an active volcano and the abundant marine life that had inspired Charles Darwin.

It gave me a lift to be going back to such a mystical place. On a previous trip, I had witnessed a feeding frenzy that was one of the most amazing sights I have ever seen. Hundreds of blue-footed boobies swept in out of nowhere to devour a school of fish that had crowded into a giant "bait ball" in a futile effort at self-protection. As they dove, the boobies folded their wings behind them, becoming lethal darts as they shot into the water. Each time one caught a fish, it popped back up to the surface, swallowed its prey, and skimmed across the water to avoid being hit by the next wave of dive-bombers. Those were some of the most vivid and enchanting experiences of my life, and I looked forward to returning.

Two of my favorite scientists joined our JASON Project team— Haraldur Sigurdsson, an Icelandic volcano expert from my alma mater, Rhode Island's Graduate School of Oceanography, and

Jerry Wellington, a delightful free spirit and an expert on the ecology of coral reefs.

One day we anchored off an island. We were going ashore to check out the iguana population, or something like that, but the others seemed to be taking forever to get ready. It didn't look that far to me, and I was impatient, so I just jumped in the water. I was wearing a white cotton turtleneck to cover a sunburn. No swim fins. I didn't realize there was a strong current. The shirt absorbed an amazing amount of water. I got about halfway to shore before I started struggling. The current was sweeping me away from the island, and my waterlogged clothing was dragging me under. I started waving desperately, trying to get someone's attention.

Fortunately, Barbara was watching. She alerted a member of the crew, and they jumped into the Zodiac and came roaring over, reached down, grabbed me by the turtleneck, and hoisted me into the dinghy. I don't remember who helped her. What I do remember is that I was going to drown, and she was there to save me.

On another day, Haraldur led us on a hike to the 5,000-foot summit of Fernandina, an active volcanic island. The footing on the lava was tough. By the time we had reached the top, we were exhausted. We barely had time to set up camp along the four-mile-wide rim of the massive volcano before it got dark.

Then, in the middle of the night, we heard such a thunderous roar that we dared not move. In the morning, we realized why it had been so loud. Our sleeping bags had been a couple hundred feet away from a major section of the rim that had collapsed into the caldera. Now only 50 to 60 feet separated us from falling in.

That would've been a bad night at the office.

—

WITH SO MUCH ADVENTURE in the air, I began to look at Barbara differently. When someone saves your life, it registers.

Maybe she was the kind of person I needed in my life. Joseph Campbell had written about how much better life is if you can find the right companion, your other half. Maybe she was the other half I was looking for. When we returned home, we began to date, doing our best to not tip off our colleagues to this new development.

I set up what I thought was a spectacular first date—lunch at the Navy Officers' Club at the Naval War College on Narragansett Bay in Newport, Rhode Island. It had been over 26 years since I'd gone on a date. I pulled out all the stops to make a good first impression as an officer and a gentleman. I wore my Navy summer whites, figuring I would cut a rather dashing profile. I had grown up in a military community, but Barbara had been raised in a more liberal household, and she was skeptical of the military. At some point during that first date, she looked me over and said, "Don't you think you'd be a little more comfortable if you changed out of your uniform?"

I began to feel better. I was getting my mojo back. I returned to Lake Ontario to do the JASON Project broadcasts. Students gathered at 14 sites around the United States and Canada to watch as we explored shipwrecks. We set up our broadcast gear on a barge anchored out in open water. Everything was going well until a storm rolled across the lake, and the tugboat captain said it was too dangerous to ferry us out to the barge. I told him I had nine live shows to do that day. And he said, "Hmmm . . . No, no, no, no."

So, there we were, stuck on land, with our first broadcast only two hours away. I had to come up with a work-around. I commandeered two phones in a bar nearby and called the TBS production crew in Atlanta. "Guys, I'll take your directions on one phone, and use the other as my outgoing mic." Here's how it worked: Through one phone, they told me what the footage was showing. I'd picture the scene and then, speaking into the other phone, describe what we were doing to students who were listening. I sat there for hours and did all nine shows that way.

I was flying down from Woods Hole to D.C. to see Barbara every chance I could. I was happier than I could remember being in a long time. She was classically beautiful, with short brown hair she swept out of her face and blue eyes that told me a very intelligent person was looking back at me. It was uncanny, the things we had in common. She had grown up loving the beach and had spent lots of time on the water on her dad's sailboat. In high school, she had helped train sea lions at the New England Aquarium in Boston. In those days she wanted to be an oceanographer and even dreamed of working at Woods Hole. When it came time for college, she had gone to Pitzer in Claremont, one of the Southern California towns near where I had lived as a boy.

But we came from really different backgrounds, and I realized that could make a life together pretty interesting. I came from corn-fed midwestern stock, parents and grandparents who had relied on common sense instead of a college education to get by. Barbara was from Darien, Connecticut, one of the wealthiest suburbs of New York City. She was raised in the Northeast—she's a *Mayflower* descendant, with family who went to some of the top liberal arts colleges. Her mother attended Juilliard as a harpist when she was 16, and Barbara grew up surrounded by art, music, and culture. When the Earles gather, you'd better bring some poetry to recite or a song to sing. One of Barbara's first questions for me was "Do you like to write poetry?"

Despite the pedigree, Barbara assured me, the Earle family would welcome me with open arms. Her mother, Bobbie, was salt of the earth in the best way. Her father, Harry, worked in the publishing industry, and the family had moved from New England to Madison, Wisconsin, so he could run the Banta Corporation, a prominent printing company. They'd returned to the Northeast when Harry retired.

When I met her parents, I figured I had to test the waters. That first night, we were sitting around having dinner and talking. I'm not sure what possessed me to do this, but I turned to her mother and asked, "Do you know why girls from Darien don't go to orgies?"

She responded calmly, "No, I don't know."

"It's because they don't want to write all those thank-you notes the next day."

To her great credit, Bobbie Earle laughed so hard she almost couldn't stop.

"I just wanted you to know who your daughter's dating," I said, grinning.

True to form, my first gift from Bobbie was personalized stationery—which I used to write a thank-you note.

—

I ASKED HER FATHER for her hand in marriage within the year. It felt like coming home. I was 48, and Barbara was 34. We were both old enough that we didn't need a long courtship.

Then Barbara had to tell Tim Kelly, her boss at National Geographic. We had kept things so quiet that when she mentioned our engagement, the poor guy almost fell out of his chair. Gil Grosvenor, then president of the National Geographic Society, threw us an engagement party in the Society's historic Hubbard Hall, grand columns and all. Gil toasted us with a great line: "It isn't that we are losing Barbara to Bob. We now have a mole in his camp." Well, the joke was on them, because in the long run, Barbara proved to be a fierce negotiator as we worked out deals with her old colleagues.

We also celebrated informally with Barbara's extended family, up in the Adirondacks, where close family friends of theirs have a compound reachable only by boat. We played a lot of paddle tennis, swam in the lake, kayaked, and hiked up Blue Mountain. I felt a sense of community for the first time in a long time, and I thought, *This is a hell of a family I'm joining.*

At first, Barbara hoped that I would move to D.C., but I couldn't leave Woods Hole without losing some of the support I needed for my expeditions, and my work there supported a large team with lots of mouths to feed. So Barbara decided to leave National Geographic. She moved up to Cape Cod in September and became president of the company I had created after finding *Titanic,* dedicated to producing books and TV specials. I renamed it Odyssey Enterprises, invoking Odysseus's epic voyage across the sea as he returned from the Trojan War. The mythic reference seemed fitting given the new epic journey Barbara and I were about to undertake.

When our families arrived for our wedding, in January 1991, it was West Coast meets East. My entire family, small as it was—my father, mother, sister, and brother, with his new wife, Jan—all flew out from California. At this point, Richard was working on his own, developing software, and although he had some absolutely brilliant concepts, he was not having much commercial success. We were all still in awe of his intelligence, and I heard my dad say to one of my in-laws-to-be, "If you really want to meet a smart dog, you should meet his brother." Dougie, who by now was also a member of the Earle clan, and my best friends from Woods Hole also were there, along with the Lehmans and Admiral Griffith and his wife. It was a wedding to remember. There had been a snowstorm, and as Barbara walked into the old stone chapel, she had on snow boots under her wedding dress. She was totally matter-of-fact about it.

For our honeymoon, Barbara and I set off to French Polynesia, because it had the most beautiful water I had ever seen. We started out at a resort and then hopped among the islands on a small cruise ship. We stopped off in California on our way home to Massachusetts so Barbara could meet my old Visigoth buddies from the YMCA high school club, who remain friends to this day.

Within days of returning home, we were working together on plans for our first expeditions as Odyssey Enterprises. Figuring out how to find shipwrecks was my part of the outfit. Creating detailed budgets and handling production details was Barbara's. It felt right. We were ready to take on the world, and I could feel the fire burning in my belly once again.

CHAPTER 10

A NEW PARTNERSHIP

Eager to throw ourselves into the thrill of discovery, Barbara and I traveled to the South Pacific in the fall of 1991. Our new production company had entered into a multiyear deal with National Geographic Television to create shows exploring the warships lost in the Battle of Guadalcanal and the wreckage of *Lusitania,* a luxury cruise liner the Germans had torpedoed during the First World War. But before Geographic would fully fund the project, they wanted us to conduct a scouting trip on a shoestring budget to dig through the mass of indistinct wrecks and see if we could identify the most important ones. And meanwhile, I was scheduled to return to the Galápagos for Jason III, the interactive shows sent live to our growing network of downlink sites.

As our plane touched down at Henderson Field in Honiara, capital of the Solomon Islands, I looked over at Barbara, who was staring pensively out the window.

"Your uncle didn't make this one," I said.

It was here that her uncle, John Earle, should have landed in 1944, in an Army transport plane on his way to New Zealand for some well-deserved rest before returning to combat. But he never arrived. Though his wife had corresponded with the War Department to try to determine where his plane went down, nothing ever came of her efforts. Barbara said she could only think about what that desolate strip of land was like in the 1940s, with the mud, rain, and malaria.

The Battle of Guadalcanal was a bloody military campaign carried out in 1942 and 1943. It involved numerous brutal battles in the jungles of Guadalcanal Island, one of the Solomon Islands, and the surrounding sea—a harrowing time for the Allies. As we taxied along the airstrip, I could see how the dense and foreboding landscape must have made everything seem more hopeless before the Allies finally wrested control of the region.

We gathered up our luggage and took a taxi to our hotel on the shores of Iron Bottom Sound, appropriately named for the numerous American, Australian, and Japanese vessels that had sunk there. We met up with members of our team, had a couple of quick beers, and headed down to the dock to board our island-hopper, a small Australian supply boat called *Restless M*.

I had suggested making this expedition Odyssey's first because I thought it would be fairly quick and easy. As a fledgling production company, we wanted to keep the stakes low and the interest high. Barbara and I might have had the trust of all our friends at National Geographic, but we still had to prove we were up to scratch if we wanted to co-produce major TV shows with them.

To save us money, Admiral Griffith let us use some of the Navy's underwater assets, including a towed side-scan sonar system and one of their deep-diving submersibles. The Navy was happy to help because it was a way to recognize the 50th anniversary of the Guadalcanal campaign. But before we could take advantage of the Navy's technical support, we had to survey Iron Bottom Sound from the *Restless M*, towing a much less sophisticated sonar system and a primitive drop camera to document the condition of at least a few of the warships. Picking through dozens of sunken ships in a relatively narrow stretch of water was a different animal from looking for *Bismarck* or *Titanic* in the open ocean—a tamer animal, and one I felt sure I could break. Plus, we had over 50 warships from which to choose.

This also was the first time that Barbara joined an oceanographic expedition. She loved being at sea; she was game for anything.

I assigned her to log data, marking down any target that resembled a shipwreck, and I put her on the midnight to 6 a.m. watch, even though she is not a late-night person. Her body adjusted to it after a sleepless day or two. She also served as navigator for that watch, and in true American fashion, a healthy competition broke out between Barbara's team and the other watch team to see which one could log the most ship contacts. She became friends with Lt. Cmdr. Chris Raney, a bright and engaging Navy officer, and developed a deeper appreciation for the military after having grown up holding that world at arm's length.

I started to think that Barbara and I could be a dynamic exploration duo, like Osa and Martin Johnson, a husband-and-wife team who had been heroes of mine growing up. They dazzled audiences in the 1920s and '30s with books and silent films on their explorations in Africa and the South Pacific. Their heyday was before my time, but I saw some of their films replayed on TV when I was a kid. And they were originally from Kansas, just like I am. I fancied Barbara and me an updated version of that famous duo, ready to explore the world together.

After all my successes and personal upheavals, the highs and lows, I figured I was overdue for a break, a chance to show Barbara just how much fun exploring could be. But in the next three years, from Guadalcanal to Galápagos to *Lusitania,* I hit the longest stretch of mishaps and bad luck ever. I've always prided myself on my ability to roll with the punches, to persevere by hook or crook. But every one of these trips made me wonder: *Can we really pull this off?*

—

OUR TEAM ON *RESTLESS M* did not have a good first day. We were testing the sonar when the ship jerked violently and the tow cable got hung up on something—it was never clear what, but it was big enough to turn the boat all the way around. It took three days to fix the cable, a waste of valuable search time. Then, when one of the mates on the bridge misunderstood our directions—thought we'd said go right when we'd said left—we ran aground. So we had to work ourselves off the bottom, and that set us back again, making our lives even harder. We were like the Keystone Cops, one stupid mishap after another.

But then we began finding targets: two airplanes and something that looked like a destroyer, surrounded by debris near the base of the volcano on Savo Island. Over the next several days, we also made a run past the Australian heavy cruiser, H.M.S. *Canberra,* one of the largest ships sunk there. U.S.S. *Atlanta,* a lighter cruiser, was harder to find, but we finally located her resting against a steep cliff in relatively shallow water.

Barbara's watch team, led by my longtime Woods Hole colleague, Andy Bowen, had originally given itself a grandiose name: Bathymetric Acquisition and Target Tracking Legion of Experts, or BATTLE. But after Commander Raney's team, the Savo Seekers, found most of the initial ships, everyone chuckled when we changed the meaning of the acronym to Below Average Target Trackers with Little Experience.

Just as we seemed to be settling into a good routine, smoke started pouring out of the generator that ran our winch, and we had to reel in the sonar equipment by hand. We were past the halfway point of the time we had for the expedition, but we still didn't have any color stills or video images of the ships we had found.

We limped back into port and headed to our hotel in Honiara.

For the next three and a half days, we waited for a local electrician to fix the generator. I was buzzing with impatient energy, and not even watching *The Godfather Part III* at the hotel could assuage my frustration. "Losing valuable time," I wrote in my diary. "Could have found two to three more ships. Damn it."

The hotel was fairly primitive, but it had a swimming pool and air-conditioned rooms, which let us escape the heat and humidity and catch up on our sleep.

Sort of.

During one of our first nights there, Barbara's screams woke me up. I jumped up and frantically scanned the room. All I could see was my shaken wife.

She just pointed at the headrest above the bed.

She had stashed a half-eaten energy bar on the ledge above her pillow, and now a giant cockroach was going to town on it. She'd opened her eyes and seen this massive, mutant thing three inches from her face. My memory is that Barbara absolutely lost it, but she says she would never have lost her cool over a cockroach.

Another day, Barbara was out by the pool, reading a book, when she heard bloodcurdling screeches. She peered over the fence, where two locals were hoisting a live pig, about to slit its throat for a barbecue.

With less than a week before we had to leave the Solomons, we decided to walk down to the warehouse where our generator was supposedly being rebuilt. We were already worried, because we kept seeing the local contractor hired to do the work dozing at one or another of the island bars. When we got to the warehouse, we found three locals literally hand-wrapping copper wire coils around the generator. They thought that was the way to fix it—but we knew there was no way it was going to work. Then, when we got the generator back, we found they had installed the starter motor backward, so it spun in the wrong direction.

When the contractor finally got everything squared away and turned on the generator, it was so out of balance it made the entire ship tremble. Then the captain fired up the main engine, and smoke began pouring out of the bridge. The smell was so noxious that Barbara and I jumped overboard into shark-infested waters to get out of the fumes.

Crew morale hit a new low. You could see it on everyone's faces. We'd found eight warships, but we still had no visual evidence. National Geographic wanted to set up a press conference to announce our discoveries. It reminded me of our *Titanic* expedition, when everyone was calling for visual proof. Sonar images are one thing, but a clear, color image of a ship is entirely different.

We had one day left before Barbara and I were scheduled to fly home. Time was running out, and I felt a little like Dorothy in *The Wizard of Oz,* watching sand run through the hourglass and marking how much time she had left to live, with the witch cackling in the background. It was now or never.

We finally got under way at midnight and headed into the sound, determined to make this work. At first light, we would go for it, stalling the ship over each of the wrecks so we could lower the cameras close enough for a good image or video footage without getting caught in the traps of twisted metal. Many larger ships had a dynamic positioning system, with bow and stern thrusters that allowed precise hovering over a target. *Restless M* had no such system. But steering a boat manually to keep it hanging motionless over a selected spot on the bottom—even with the current pushing in one direction and the winds in another—was a ballet I had mastered over the years. So I asked permission from the captain and took the helm.

We positioned ourselves over one shipwreck after another, lowering the rig, turning on the cameras, and capturing images. Or at least we hoped we were. Time seemed to melt away. Around midnight, after a final run over *Canberra*, we pulled the camera back aboard for the last time and gathered in the control van to view the results. Cheers went up as we saw images of antiaircraft guns. We had done it. We had pulled the expedition out of the fire with only hours to go.

As *Restless M* headed back into port, everyone tumbled into bunks and went to sleep, satisfied. Too soon, we were awakened abruptly. *Restless M* needed to move out to make room for another vessel, and Barbara and I had to disembark immediately if we wanted to catch our flight back home. We watched from the shore as *Restless M* pulled away, leaving us no chance to say goodbye to the crew members who had gone through so much with us.

We had managed to find eight wrecks, and after Barbara and I left, the team found two more—enough to get us the green light for a better-funded return trip in 1992. The Navy, through Submarine Development Group One, lent us *Scorpio*, a deep-towed side-scan sonar system, and *Sea Cliff*, a sister submersible to *Alvin*. We found three more wrecks and shot amazing video for the television special, *The Lost Fleet of Guadalcanal*, featuring close-up views of famed wrecks such as Australia's *Canberra* and America's *Quincy*, both sunk during World War II—a part of history that had been long lost.

Barbara and I were proving a great team, but she informed me of one important adjustment if we were going to keep exploring together. During one of our outings in the Solomons, I had asked Barbara to say something on camera. She froze as soon as the film started rolling. Only a couple words came out of her mouth. Then she started laughing. "Bob, you do what you do best," she said, "and I'll do what I do best—and that's to be behind the camera. You are the king of the 30-second sound bite."

—

A MONTH AFTER our first Guadalcanal trip, I was back in the Galápagos for Jason III, out in a small dive boat with a National Geographic television crew headed by my friend Chris Weber. I had been in and out of the water all day, and I was freezing as I slipped out of my wet suit. I also was a bit shaken. We had been filming a group of female sea lions playing underwater, and all of a sudden an oversexed male sea lion had charged me. I must have come too close to his harem, putting him on the offensive. He charged straight at me, full speed. Thank goodness he turned away at the last second. His one-ton body could have easily killed me.

While we were filming the sea life, the Ecuadorian navy was towing a barge loaded with six million dollars' worth of our equipment to the islands. The Galápagos National Park Service had required us to set up our production operation on a barge rather than on land, to protect the island's fragile ecosystem. Thanks to Barbara's diplomacy, top Ecuadorian officials had promised to get the barge there for us, and it was carrying our satellite gear, a portable production center, generators, multiple color camera systems, and the two *Jason Jr.* robots we now had.

As I stood there, shivering in a light breeze, a call came through on our radio. The voice was barely audible and badly broken up.

"*Nortada,* this is Puerto Ayora, on the island of Isla Santa Cruz. This is an emergency, please come in."

Chris picked up the handset. "Puerto Ayora, this is the *Nortada,* over."

It was Andy Bowen, who was handling all the technical aspects of the expedition. He asked to speak to me. The static got worse.

"Andy, this is Bob, what's up? Over."

I could hardly hear his response as a chopped-up series of words came in.

"The barge—"

"What about the barge?"

More static, and the only other word I could make out was "sank."

"Did you say the barge sank?"

Andy's defeated "yes" cut through the white noise.

Oh, my God. *Not again,* I thought, as I stared down at the sea around me.

What was it about the JASON Project, I wondered, that seemed to always bring on disaster? Just two years before, *Jason,* newly operational, had plunged to the sea bottom days before our first live linkup. We had been able to retrieve it, but this sounded worse. Now two *Jason Jr.* vehicles were dead on the bottom of the ocean. I messaged the five charter boats carrying our various work teams to meet us the next day off Baltra Island, near the center of the Galápagos chain, where the airport is located. The next morning, they were all there at anchor, bobbing up and down on the swell. Bad news travels fast, and there were a lot of sober faces.

We began the difficult job of determining how we were going to work a miracle. I tapped a team, led by Cathy Offinger, to fly with me to Quito, Ecuador's capital, where we could use a bank of phones to conduct what amounted to a giant scavenger hunt. The wish list to replace what we had lost in several 20-foot shipping containers must have been 10 feet long, and our pitch was basically the same no matter who we called: *Remember that generator you sold us? Well, it's at the bottom of the ocean, in 9,000 feet of water. We need to borrow another one, and we don't have any money. What do you say?*

It was amazing how many people were saying yes.

Piece by piece, we rebuilt the mission. We rented a warehouse near the Miami airport for all our gear. Now we had to figure out how we were going to get the stuff to Quito and on to Baltra Island.

Tim Armor, then the executive director of the JASON Foundation, knew a guy in the Air Force who knew a guy in the CIA, which was operating an air transport company out of Miami. The company agreed to make a touch-and-go run for us, dropping everything on the tarmac in Quito.

For the next step of this delicate operation, I needed a secret weapon: my wife. Barbara had experience dealing with foreign officials from her Geographic days, but she was back home in Cape Cod. I asked her to catch the next plane to Quito and meet with one of the Ecuadorian officials we'd gotten to know on our scouting trip the year before. Ecuador had a cargo plane large enough to hold our gear but small enough to land on Baltra's short gravel runway. She needed to get it for us.

She arrived the next day and went straight to the business dinner. I spent the evening pacing back and forth, waiting anxiously for the result. Finally, I heard the key opening the door.

"You got your plane," Barbara said.

We were ready to go.

Right after the cargo plane took off to deliver our new generator and transmission gear, Barbara and I flew back to the States. Without our production van, it would be impossible for me to host two weeks of live broadcasts from the Galápagos. I had to do them from CNN's studios in Atlanta, much like the call-ins I improvised from a barstool at Lake Ontario. Barbara, all too familiar with my fly-by-the-seat-of-your-pants MO, headed back to the Cape while I checked into a hotel along I-85, a far cry from the on-location excitement I had envisioned from the Galápagos.

My improvised studio in Atlanta was a small editing suite, complete with an oblivious intern sitting at a computer screen with a headset on, chewing gum and dancing in her chair. The cameraman set up in the hall and had to zoom in just enough to avoid the intern. As long as she didn't lean forward, we were fine.

I had a small palm tree next to me and we projected pictures of animals in the Galápagos to make it look like I was on location.

We willed it to work, and it did.

Jerry Wellington, a dear friend, stole the show with his segment on how tiny damselfish create their own algae gardens and then fight to protect the gardens from intruders. Jerry wore a large, round diving helmet with a microphone, air hose, and communications cable attached, and the kids knew him as Dr. Bubblehead. I returned to the Galápagos 24 years later to spread his ashes after his death from early-onset Alzheimer's disease.

Watching colleagues like Jerry bring his enthusiasm and verve to classrooms a world away made all the struggles worthwhile. And with our next *Jason* expedition, to Mexico's Sea of Cortez in 1993, we were finally able to demonstrate the value of telepresence to the academic community, beaming live video from the expedition right to their college and university offices. That trip also included the requisite misadventure, when a Cessna we rented broke down, stranding us in a desert. Thank God the radio still worked, and the pilot was able to call in another plane to pick us up. Through it all, the JASON Project showed that we could win the hearts and minds of thousands of schoolchildren this way.

—

IN THE SUMMER OF 1993, Barbara and I were off on our next adventure together, flying to Ireland to explore the wreckage of *Lusitania,* a luxury cruise liner that had been a rival of *Titanic.* I'm always in a hurry, and I tend to walk quickly, while Barbara likes to take her time. Airports really bring it out in me. The first time we were ever together at an airport, Barbara told me later, she made a conscious decision to see how I would handle it if she just continued along without hurrying while I rushed ahead. I circled back, walked alongside her as long as I could, then my faster pace pushed me ahead of her. I kept going and circled back again. When I think about it, that pattern—her patient trek and my blasting ahead, full speed—reflects a dynamic of our relationship, a give-and-take we had begun to explore in our time together.

At the same time, we were working on Steven Spielberg's TV show *seaQuest DSV,* science fiction set under the sea. I proofread the scripts for technical accuracy and did a 90-second spot at the end of each episode, explaining the science on which it had been based. Barbara filmed me, mostly on the beach in Cape Cod, and produced my segments.

Sometimes we had to laugh at what it took to come up with a believable explanation for the episodes. Roy Scheider played a submarine captain, and in one show his crew came across a sunken ship that still had people living inside, basically like in *The Poseidon Adventure.* The only thing I could come up with was that the people they encountered weren't really people, but figments of their imaginations brought on by nitrogen narcosis. It was a stretch, but it worked.

We had something else to think about—and celebrate—as we set off for *Lusitania.* Barbara was pregnant. Both of us were looking forward to a new baby, and I was excited, despite my recent grief that had felt all-consuming. With Dougie in his early 20s, it had been a long time since I had held an infant and witnessed all the thrilling firsts—the gurgling laughter, the determined army crawl across toy-strewn floors, the first toddling, cautious steps into a parent's loving arms. The upcoming birth rejuvenated and inspired me.

We positioned *Northern Horizon,* our command ship, 11 miles off Ireland's southern coast. We knew where *Lusitania* lay—that wasn't the issue. Our goal was to investigate the mystery of what had sent the grand ship to her watery grave, taking the lives of about 1,200 passengers and crew members in May 1915. We knew the ship had sustained two explosions. The Germans had proudly admitted to firing a torpedo into her starboard bow—the first explosion—and justified it by claiming that the passenger liner was carrying illegal war materials. The second and fatal explosion was the mystery. The Germans said those very munitions had erupted after the torpedo struck the ship's forward magazine, but the British government denied that claim. The ship was carrying munitions, they granted, but only small-arms cartridges without gunpowder—legal cargo that would not have caused an explosion. They maintained that position for decades, but in 1982, the British government had warned divers that a large amount of combustible ammunition was still in the wreck.

I wanted to get a look at the area near that forward magazine, but I wasn't sure if that would be possible, because *Lusitania* was resting on her starboard side. The wreckage also had become a favorite place for local fishermen to set their nets, and some of the nets had broken off and gotten caught on the ship's superstructure, forming a giant spider's web that could endanger our manned mini-sub *Delta,* our newly developed *Jason* robot, and a smaller remote-controlled vehicle called *Homer.*

The first step was to use *Jason* and its sophisticated sonars to make a 3-D model of the wreck. We could then superimpose the ship's original engineering drawings onto that map to locate the forward magazine precisely.

Fortunately, the angle of *Lusitania*'s hull made it possible for *Homer* to slip under the bow and determine that the forward magazine was completely intact. Whatever had caused the second explosion had not been stored there.

We kept exploring very cautiously. *Lusitania* was only about 300 feet down, and although you would think a shallow shipwreck was safer, I have always found those in shallower water to be among the most dangerous dives. My fears were confirmed when we got a call from Chris James, our *Delta* pilot. He was driving maritime artist Ken Marschall and maritime historian Eric Sauder down to get a close look at the shipwreck.

"The propeller has sucked in a fishing net," he said. "I can't move."

I could just imagine it—snarls of old fishing net wrapped around and around the propeller shaft, making it impossible for the prop to spin. They had lost their ability to move in any direction through the water.

When Chris tried to kick the sub up to top speed to pull out of the net, the lights dimmed, power output dropped, and forward progress came to an abrupt halt. There was no way the sub could power itself out of this situation. The only option left was to unscrew the propeller assembly from inside the sub. It would fall away, and a rubber gasket was supposed to keep the seawater from flooding into the shaft opening. The sub couldn't power itself anymore, but it was buoyant and would rise to the surface on its own once freed.

I'd never lost anyone on an expedition, and I didn't intend to start now. As Barbara and the rest of my team looked on, a hush filled the control center. Then we heard a soft voice coming through the sound system. "Righty tighty, lefty loosey"—the jingle everyone uses to turn a screw in the right direction. "*Northern Horizon,* we are free. Coming up!"

But that was not our final encounter with *Lusitania*'s traps. Chris and the others on *Delta* had noticed big chunks of coal lying on the bottom. They must have spilled out of the side of *Lusitania* as she had plunged forward on her short trip down. I sent the *Jason* robot down to use its mechanical arm to recover some of that coal for analysis ashore.

Cyril Spurr, a British munitions expert who was with us, had noted that if the German torpedo had hit behind the forward magazine, it would have struck a coal bunker. Because *Lusitania* was nearing the completion of her long voyage across the North Atlantic, the large bunker would have been nearly empty, filled with a thick layer of warm coal dust. If the torpedo had violently pushed that coal dust up into the well-oxygenated air inside the bunker, a spark could have ignited and set off the second explosion.

Jason grabbed the coal just as planned, but as it was coming back up to the surface, it tangled up in the same nets that had trapped *Delta*. No humans were at risk this time, but *Jason* had cost over a million dollars to build, and we wanted our valuable asset back safe and sound.

What ensued was a tug-of-war between our ship and the shipwreck below. We could reel *Jason* up to about 130 feet—shallow enough to reach with divers. Two members of our team—National Geographic underwater photographer Mark Shelley and consultant Paul Matthias, who led our 3-D modeling effort—volunteered to dive down to cut *Jason* free.

Sending down two humans to save her scared the hell out of me. I was terrified of losing people on my watch, and Todd's death made me all the warier.

We pulled up *Jason*'s tether as far as we could and held station above the wreck while our team disconnected *Jason* from the ship and attached it to buoys. Then we backed the ship away, watching as Mark and Paul donned their wet suits and tanks, climbed into our small Zodiac, and drove over to the floats bobbing up and down in the swells. My heart was in my throat as they both disappeared between the cold, green waves. Seconds turned into minutes, and then tens of minutes.

Finally, two heads popped up, then two hands with "thumbs-up" signals. We had stolen success out of the jaws of failure. My people, and my gear, were safe after all. We also had determined that the coal dust could have helped trigger the second explosion on *Lusitania*—and now had compelling evidence for historians to analyze.

—

WE FINISHED THE EXPEDITION just in time for Barbara and me to drive up to Northern Ireland to attend the wedding of Hagen Schempf, who had delivered the beautiful eulogy for Todd. We stayed at home for the last few months of Barbara's pregnancy, and in January 1994, Benjamin was born—Barbara's first child and my third.

What had begun as a working relationship had blossomed into so much more. It was a joy to be married to someone so generous and giving, so brilliant and kind. I had at last found someone who understood me and would continue to work by my side—the other half of my soul, as Joseph Campbell expressed it. I knew Ben would have a mother who would teach him not only to love, but also to ask questions and dream big. I no longer felt isolated or trapped. I had a companion I could walk side by side with for the rest of my life.

CHAPTER 11

ROLLING THE DICE

I n the spring of 1994, I received a phone call out of the blue. It was from Hugh Connell, a retired lawyer and the president of the Mystic Marinelife Aquarium in Connecticut. He had courted me a couple of years before with his big plans for the aquarium. It was a major economic driver for the southeastern region of the state, and he wanted to raise tens of millions of dollars to renovate and expand it. He had visited me at Woods Hole and convinced me to join the aquarium's board to help raise its visibility.

Now Hugh was calling to say that Connecticut's governor, Lowell Weicker, wanted to see us the following day. I asked why, and Hugh said he supposed his request for state funding for the expansion plan had reached the governor's desk.

The next day, Hugh picked me up in his small private plane, and we flew from Hyannis to Hartford. When we got to the governor's ornate office, Weicker—all six feet six inches of him—stood up to greet us. The man was a certified Connecticut institution, a former congressman and U.S. senator who had been one of the first Senate Republicans to call for Richard Nixon's resignation in the Watergate scandal. He towered over us, holding the proposal that I thought we were there to discuss, then he slammed it down on the desk in front of him.

"Dead on arrival," he said to Hugh. "Unless you can get Ballard to move to Connecticut."

I sat there, stunned. I had joined the aquarium board as a favor to Hugh, so he could use my name and connection to the sea in his fundraising. But I certainly didn't expect this—that the governor of Connecticut would be demanding I leave Woods Hole and join the Mystic operation, or else he would dash Hugh's dream.

We just listened. I was still in shock as we drove back to the airport, but Hugh didn't seem nearly as alarmed as I was. In fact, he seemed almost casual about the proposal.

"So, what's it gonna take to get you to do it?" he asked.

"I don't know," I said. "I've got to think about it. I mean, I'm totally caught by surprise on this."

As shocked as I was, I wasn't going to dismiss the idea out of hand. Maybe there was an opportunity worth considering here. Leaving Woods Hole would be risky in many ways. I'd lose access to the magnificent team of engineers and undersea experts I'd nurtured and the incredible remotely controlled vehicles and support systems we'd developed. I'd also lose the prestige of being associated with one of the world's top oceanographic institutions. I'd literally be starting over.

But my emotional connection to Woods Hole had been waning for a long time. I'd been struggling to live within its strictures ever since I'd found *Titanic* in 1985. The institution's leaders had always had a love-hate relationship with that discovery. It was the key to Woods Hole's fame, and some of its fundraising successes, but it still rubbed key people the wrong way. And everything I'd done since—starting the JASON Project; finding *Isis* and *Bismarck,* the World War II ships, and *Lusitania;* my books,

National Geographic shows and articles—had just added to the tensions.

I understood where they were coming from—hard-core scientific research was Woods Hole's DNA. John Steele had been replaced by Craig Dorman as Woods Hole's director, and then Dorman had been replaced by Robert Gagosian, and with each step in the succession there was growing friction over my interest in popularizing science, not to mention irritation that they couldn't control me. I don't like to be fenced in—that's why I don't want to have a single master—and if someone tries to corral me, I tend to just slip out and move on.

The more I thought about it, the more I realized I had changed, but Woods Hole hadn't. I'd outgrown the place.

I'd already been fishing around a bit. I had been looking at whether I could go back to Stanford, where I'd had such a productive sabbatical dreaming up my robotic camera systems. I'd also touched base with my perennial rival, Dr. Spiess, at Scripps. Now, 30 years after rejecting me for its graduate program, he was interested in having me go to work there, which was immensely gratifying. It felt like a measure of redemption.

Barbara wanted to stay in New England, given her family's deep roots in Connecticut, and that sure did narrow the field. I called to tell her what Weicker had said, and as soon as the words "move to Connecticut" came out of my mouth, she was all in.

Just like that, my new path was clear.

It was risky to start a new venture, but I'd spent my life taking risks. I run the numbers, assess the odds. I know that if you don't stick your neck out, you're just going to be left sitting there with the rest of the pack. I take calculated—not reckless—risks, but I do roll the dice.

I began my negotiations, asking Hugh and the governor to build a new professional home for me at the aquarium complex, called the Institute for Exploration. Governor Weicker made it happen. In late November, I spent a day at the state capitol in Hartford, answering last-minute questions from lawmakers as they voted to approve $30 million in state assistance to expand the aquarium. I agreed that I would stay on for at least 15 years, to help ensure that the deal paid off for the state. The Sea Research Foundation, which owned the complex, and I agreed to raise an additional $15 million from donors.

It was going to take several years. We would design and construct a 25,000-square-foot building to house my institute and the JASON Project, along with an exhibit center called "Challenge of the Deep." That hall would have a permanent *Titanic* exhibit and displays related to my other expeditions. While construction was going on, I would stay at Woods Hole and consult part-time in Mystic. I set my 55th birthday, June 30, 1997, as my last day at Woods Hole. I was still recovering financially from my divorce, and if I waited, I could leave with better medical and retirement benefits.

I also needed time to take care of my Woods Hole team, and that's one reason why officials there were willing to go along with the lengthy transition. Dana Yoerger, Cathy Offinger, Andy Bowen, and many others had sacrificed so much for me and my projects, and I wanted to make sure their futures were secure. I'm sure that Gagosian was thrilled I was leaving, but several members of the Woods Hole board did not want me to go, and naturally my team was nervous, even though I told them I would keep using them on projects, which I do to this day.

I'm a very patient planner, so the wait didn't bother me. I was ready to take on this next stage in my life. I had figured out what I needed most, and that was the freedom to chart my own destiny. All I wanted was a license to hunt.

—

I SHOVED OFF FROM CRETE on a Navy research vessel in September 1995, the smile spreading under my baseball cap. I was thrilled to be resuming my search for ancient ships, the thing I had most wanted to do since we'd found *Titanic*. It had been seven years since we'd found the remnants of the Roman ship we named *Isis,* but one of anything does not a pattern make. I was convinced more wrecks were nearby. I was determined to prove that searching for archaeological artifacts in the deep sea could help find trade routes that would tell us more about ancient civilizations.

Up till this time, most marine archaeologists had worked close to shore, using scuba gear to explore wrecks in relatively shallow water. Many of them believed that ancient mariners had hugged the coast, fearful of losing sight of land, and worked their way around the edge of the Mediterranean Sea. Those were the ancient seaways they had been mapping. *Come on,* I thought. If anyone back then wanted to sail between Rome and Carthage, I'll bet they went in a straight line. They must have been tight with money, and eager to save time, just like modern seamen. Wouldn't some brave soul have ventured straight across and emboldened the others when he returned unharmed?

I like to joke that this was my most sophisticated search strategy. I took out a ruler and drew a line from Rome to Carthage, in present-day Tunisia. It was just midwestern common sense, like my grandmother taught me. These ships were loaded with amphoras full of wine and other goods. Surely, the sailors dipped into the wine on lonely nights at sea. It's just human nature. Wouldn't some of them have chucked the empties overboard? And when they were afraid of sinking in a storm, wouldn't they have shoved some of the heavier amphoras over the side? You can even read it in the Bible. When the disciple Paul sailed from the Holy Land to Rome, he said, "We were tossed so violently that the next day the men began to jettison the cargo" (Acts 27:18).

I knew from the *Isis* wreck that the wood-eating mollusks would have devoured the hulls of Roman and Carthaginian ships long ago. But amphoras were made of clay, not wood. And sediment builds up very slowly in the deep—only about four-tenths of an inch during each millennium—and even lying on their sides, the amphoras were a foot high. You couldn't hide those suckers. I figured that this ancient trade route was going to look like I-95 would after two millennia *without* an Adopt-a-Highway program. It was going to be littered with trash, and we were going to look for the empties.

In my mind, this was about the surest bet I'd ever make.

I also was eager to show the marine archaeologists they were wrong. I had good relationships with some members of that academic community. Anna McCann had been on our *Isis* expedition, and we followed her advice on which artifacts to raise and how to protect them. But others were publicly questioning whether what I was doing was really archaeology. One had even written to my government sponsors, trying to get them to cut off my funding. For those who had spent years of their careers diving onto wrecks with scuba gear, eyeing them directly, and lovingly excavating artifacts by hand, robots like *Jason* took away the romance. Maybe they were jealous, frankly. I had no formal training in archaeology. I also had raised plenty of money, and I was making discoveries.

In fact, those skeptics were probably even more perturbed when the Navy took me out on *NR-1,* the only nuclear-powered research submarine. I'd gone out on it for nearly a month in 1984, when I was visualizing how to adapt the Army's terrain warfare concepts for undersea battles. The 147-foot-long sub was based in Groton, Connecticut, and I'd gotten reacquainted when the Navy had invited me to go on a 1994 training exercise to check the wreckage of S.S. *Andrea Doria,* an Italian luxury liner that had sunk off Massachusetts in 1956.

NR-1 was ideal for hunting deepwater shipwrecks. It could dive to 3,000 feet, 300 feet deeper than where *Isis* lay, and it had sonar systems that could spot a beer can from almost two miles away. It also could stay underwater for up to a month at a time. The Navy had used the sub for underwater intelligence spying, and now that the Cold War was over, it needed work to keep its crew members sharp. I was happy to help out.

We took *NR-1* on a practice run off Greece in August 1995, to film the wreckage of the *Britannic* for a PBS show called *Titanic's Lost Sister*. After a short break, we set sail again from Crete, heading toward the *Isis* site in the Skerki Bank area, west of Sicily and north of Tunisia. Our plan was to scout for other wrecks and then return to recover artifacts in 1997, with Dr. McCann and my *Jason* robot.

Even though *NR-1* was much larger than *Alvin* and could accommodate 11 crew members and two scientists, it felt cramped. The compartments for the nuclear reactor and the engines took up so much of the vessel. But the tight spaces didn't bother me, because I had spent so much time in smaller submersibles. I loved crawling down to the bottom of the sub to look out one of the three viewports, pressing my face to the glass just inches above the ocean floor. It was like lying on a foam bed, but the area was shaped like a bowl, so you had to tighten your muscles to keep from rolling around.

In 1995, we discovered a sunken 19th-century sailing ship and the remnants of two Roman ships. Returning in 1997, we hit the mother lode. As *NR-1* drove along the ocean floor, I stayed in the command center on a Navy support ship, the M/V *Carolyn Chouest*. My role was to give instructions to David Mindell, a Woods Hole engineer and search leader on the sub. The cardinal rule was "Stick to the track lines"—the systematic search plan we had laid out. In other words, don't get distracted by new targets that pop up.

An hour or so into the run, an *NR-1* officer picked up something on sonar that seemed man-made and suggested that Dave divert to check it out. Dave hesitated, worried about what I might say, and then decided to trust the officer's expertise. So much for discipline—but wouldn't you know that this turned out to be the biggest find of the cruise?

As Dave told the story later in his book *Our Robots, Ourselves,* he looked out the viewport into the green glow of *NR-1*'s lights, and saw more than 100 ancient ceramic jars lying on the seabed in two clusters about 30 feet apart, exactly where they must have been stacked inside a wooden hull with two cargo holds before it rotted away. Parts of two lead anchors showed where a ship's bow had been.

Next we sent down *Jason,* coming closer than 10 feet above the wreck, taking overlapping photos and sonar images to compile into a digital photomosaic, every detail captured but the remnants of the ship entirely untouched. Dave actually thought he got a much deeper understanding of the wreck from studying this digital data than from seeing it himself down on the seabed.

We printed out the mosaic, and I spread it across the chart table in the control van. Anna McCann and I leaned over it together. "Which ones do you want?" I asked her.

"That one, that one, that one, and that one," she said, pointing to different artifacts.

Then we put our one-and-a-half-ton yellow piece of equipment back to work. *Jason* hovered just above the artifacts and scooped them up gently with mesh gloves as we watched on computer screens, the robot's hand and arm movements nearly hidden by swirls of sediment as it deposited each object on the elevator's mesh floor.

McCann selected 35 artifacts, including a bronze ladle and saucepan, ceramic cooking pots and jars, and six types of amphoras. We brought them up two to four at a time and wrapped them in gauze. They all survived without a scratch. Looking at the array of artifacts, we began to understand that this ship, which we dubbed *Skerki D,* was like an ancient Walmart, considering the amazing variety of stuff it had on board. It was the largest Roman ship we had found. Once she analyzed everything, Anna determined that it was the oldest one we had found, too, most likely dating back to between 80 and 60 B.C.

In less than a week, we located two more Roman wrecks and retrieved 115 artifacts from the ships we had found in the two expeditions. All told, we found the largest concentration of ancient shipwrecks ever discovered in deep water, dating from about 100 B.C. to A.D. 400. It was quite a bonanza. And there it was, the trade route. Now, did I have to have a degree in archaeology to figure that out?

—

SOMEONE WAS MISSING on these Skerki Bank runs—Barbara, the other half of our modern-day explorer couple. After Ben was born, she had insisted that we not be absentee parents. Someone needs to be home, Barbara said, and she felt it made the most sense for her to be the one. I needed to go to sea, and she had the flexibility to work from her office at home. I agreed to limit the length of my cruises. I wanted to spend more time at home with my family than I'd been able to do when chasing my Ph.D. and tenure and spending so much time at sea.

Some might think I was too deeply immersed in my work to volunteer to stay home, and I wouldn't argue with that. But Barbara also noted that as the front man of our various enterprises, I needed to remain visible, while she had plenty to do behind the scenes. Besides planning new projects for our production company—like a search for U.S.S. *Yorktown,* the World War II aircraft carrier—she was doing consulting work to help set up the institute in Mystic. I remember her saying she wanted to give me the freedom to keep dreaming—an incredible gift on her part.

It had been clear from the start of our relationship that we were wired differently. I often joked that I looked at the big picture and hated the details, while Barbara was totally detail-oriented—a talent that made her a great TV producer. As we settled into our marriage, though, I could see that Barbara had a more nuanced view of our differences. I might spin off, totally absorbed in the project of the moment, and thankfully, she realized it wasn't that I was being selfish. I just couldn't turn off the torrent of ideas rushing through my brain. It sometimes left her breathless, she said, how consumed I was with ideas, one coming right after another. She marveled at how easily I could put disparate thoughts or people together and find ways to make them work, like persuading the Navy to let me use *NR-1* to take a turn at archaeology, for example. "I'd be exhausted if I was in your head," she says to this day.

We complemented each other well, but Barbara began to realize that she could benefit from an occasional break from my tornado, too. Those are the quiet times, she says, when she can sit back, take a breath, and reestablish her own equilibrium. The occasional separation has helped us maintain a good foundation. I knew she always had my back, even when I was out at sea. Whatever happened with Ben, our home, or our business, I knew she could handle it.

I kept the Skerki Bank trips to one to two weeks, no longer, and I didn't go to sea at all in 1996. But I was on the road a lot, giving 20 to 30 lectures each year on college campuses, talking about my projects and rebuilding my finances. I was also busy raising some of the $15 million we needed to finish the institute at Mystic. And I was going on fun JASON Project trips each year, taking students with me in person and via telepresence.

Barbara and little Ben came on some of my travels with me. We went to the Florida Keys, where the Navy let the *Jason* team use *NR-1* to explore an undersea sinkhole. We shared a wonderful family moment in a cottage on Hawaii's Big Island when Ben, 11 months old, pushed himself up and took his first steps.

We were in Hawaii to film one of our *Jason* shows when I had this crazy notion that each episode would open with me leaning out of a helicopter with the door removed to talk to a cameraman standing on its skids as we hovered over a molten lava flow. We headed out over the lava, and the cameraman, attached to his own safety lines, reached over to adjust my microphone. I was leaning out of the copter on my safety lines to talk into the camera when I suddenly realized that he had accidentally unhitched my safety harness. I grabbed the door frame and pushed myself back into the copter. I would have been roasted. Needless to say, that footage did not get shown to the kids back home.

To keep our family connected, Barbara also insisted we take several vacations together each year that weren't work-related—to the Bahamas, Wyoming, and Rhode Island's Block Island. She liked to pick places that had poor cell phone connections to force me to keep my work at bay.

—

BY EARLY 1997, the time had come to leave Woods Hole and move to Connecticut. We put our house up for sale, figuring we'd have time to pack gradually. The house sold immediately, however, and by March we were scrambling to find temporary housing in Connecticut.

Barbara was pregnant again—delightful news, and another test of her ability to calmly handle any and all situations. We wanted to build a house near Lyme, a town that dates back to the 1600s. Barbara's family had friends there, and it was only a half-hour drive from Mystic. Barbara arranged for us to sublet the parsonage house at a local church, and then we rented a house at Griswold Point, a gorgeous area where the Connecticut River empties into Long Island Sound. She also worked her magic to find a nice piece of wooded property where we could build a house like the one we'd had on Cape Cod.

My fundraising for the Mystic complex was going well. The Kaplan Fund—a prominent New York philanthropic organization founded on money from the Welch's juice company—was one of the first donors. William Simon, the former Treasury secretary, and Harry Gray, the former chief executive of United Technologies, each pledged one million dollars, and the Mashantucket Pequot Tribal Nation, which was operating the Foxwoods Resort Casino, promised five million dollars. Construction was taking longer than we'd expected. I was meeting regularly with architect César Pelli and exhibits designer Tom Hennes, working on my "Challenge of the Deep" exhibit hall. We had decided it would feature a large model of *Titanic*.

I was talking a lot to James Cameron, too. He had already directed *Terminator* and *Aliens,* and now he was making *Titanic*. Not long after we'd found the ship, Cameron had visited me at Woods Hole. He wanted to learn all he could to dramatize the story accurately. He eventually built the story around a fictional romance between a struggling artist and a young society woman, played by Leonardo DiCaprio and Kate Winslet. Jim was obsessed with getting all the details of the ship right, and sometimes he'd call late at night from the set in Ensenada, Mexico, or the editing room in Hollywood, and he'd be all excited and ask me a bunch of technical questions.

When the movie was about to come out in December 1997, Jim invited me to attend a Hollywood screening. But our daughter, Emily, had just been born on November 26, and I told Jim I wanted to stay home with my family. So then he invited me to a screening in Washington, D.C., and I said I could hop down for that one. When he knew I would be there, Jim arranged for Larry King, the CNN talk show host, to interview us together.

Sure, I thought. *Sounds fun.* But there was just one problem: The interview was scheduled to take place before I had seen the movie.

So Jim found a cooperative theater 30 miles from our house and sent a limo to carry me, Barbara, and not-quite-three-week-old Emily to an 8 a.m. showing of *Titanic*. Barbara was so exhausted from lack of sleep that she had the limo stop at a Dunkin' Donuts to buy some coffee. Then we settled down in the center seats of this giant theater, just the three of us.

It's a long movie. Halfway through, Barbara had to go to the bathroom. I stood up and said, "Stop the movie!" Can't do that every day.

The next day I flew to D.C. First, we taped the *Larry King Show*. Then we attended a party at Vice President Al Gore's home at the Naval Observatory. As the party wound down, Jim grabbed my arm and said we were going to ride to the theater together.

We sat in our limo and watched all the others pull out. We were still just sitting there. Everyone else was gone. I glanced at my watch and asked why we were just sitting in the vice president's parking lot.

"We're waiting to make the biggest splash," Jim said.

Finally, we pulled up to the front of the theater. It was a mad-house, with dozens of flashing cameras and people lining both sides of the red carpet. Jim turned to me and grinned. "You go first," he said. "You found it."

Oh, OK. Straightening my lapels and smoothing my shirt, I opened the door and climbed out of the limo, plunging into the flashbulbs and cacophony.

All the photographers glanced at one another, murmuring back and forth in confusion. *Who is this guy?*

Then, just as quickly, a frisson rippled through the crowd. *Oh, he's the guy who found* Titanic! All at once they started yelling and screaming like it was the Oscars or something.

I posed for pictures and went into the theater while Jim made his entrance.

I loved the movie. It was a wonderful love story, with that scene of Winslet and DiCaprio standing on the prow as she spreads her arms wide and cries, "I'm flying," right before their first kiss. It was such a beautiful and meticulous re-creation of the ship, too, from the Grand Staircase to the crow's nest. I knew what the old lady looked like in her grave, and Jim showed me what she'd looked like as a young lady when she'd sailed from England.

Cameron's movie also captured the random nature of who lived and who died. I must say, that's a question that has stuck with me ever since my summer in Army boot camp. Barbara's father, Harry, was shaken by that reality when he saw the movie, too. It reminded him of flying missions in World War II and seeing nearby planes vanish, hit by German flak. Who died? Who survived? For him, as on *Titanic,* it was a toss-up.

—

PARTLY TO HONOR the World War II generation, my first search on behalf of our new Institute for Exploration focused on U.S.S. *Yorktown,* one of three American aircraft carriers in the Battle of Midway, the turning point in the war in the Pacific. *Yorktown* had limped into Pearl Harbor shortly before the battle, trailing a 10-mile-long oil slick, the result of severe bomb damage from Japanese planes in the Battle of the Coral Sea. Shipyard engineers had said repairs would take weeks, but Adm. Chester Nimitz, commander in chief of the Pacific Fleet, said he needed the ship in three days—and he got it. With the help of superlative code breaking by his intelligence team, Nimitz knew that the Japanese were sending a huge force to try to capture Midway Island. He planned to ambush Japan's fleet, and *Yorktown* was a key part of the force. Dive-bombers flying from U.S.S. *Yorktown,* U.S.S. *Enterprise,* and U.S.S. *Hornet* crippled all four of the Japanese aircraft carriers that had attacked Pearl Harbor, and they all eventually sank. Japan's torpedo-launching planes, in turn, severely damaged *Yorktown* with two almost perfectly placed strikes, and after most of the more than 2,000 crew members had abandoned ship, a Japanese submarine sank the carrier with two more torpedoes. All told, 141 of Yorktown's crew members died in the battle.

I was feeling a strong sense of mission and urgency on behalf of the men who had lived through these horrific battles. Most of them were now in their mid-70s to mid-80s, and if those remaining *Yorktown* survivors were ever to see their ship again, now was the time to find it.

My dad was part of that Greatest Generation. Even though his eyesight had prevented him from flying warplanes, he'd contributed to the effort by helping to build B-29s at Boeing. Now, at 82, his health was failing. Doctors had botched his heart bypass surgery. He got a massive staph infection and almost died several times as they tried to stabilize him. In the process, part of his brain was destroyed, and he lost his short-term memory. My brother, Richard, summed it up best: Life had become an insult to him. He had had a great mind, and it wasn't there anymore.

Dad and I had gotten closer over the last few years. At some point when I was visiting him in California, he had said, "You must have been disappointed in me as a father." I was taken aback by that comment. Here I'd spent so much of my life feeling I hadn't measured up to his expectations, and then I hear him say he was worried that he didn't measure up to mine.

"That's not so," I responded. He had gone from being an orphan on a ranch in Montana to serving as chief engineer for part of the Minuteman missile program during the Cold War, and that was a greater accomplishment than mine.

Our relationship changed in those last years. There was one other moment I'll never forget. My sister, Nancy Ann, had always lived with my parents. Now Dad asked me to care for her once he and Mom were gone—another sign that he was pleased with what I'd made of myself. He trusted me to take over in his place.

As we were putting together plans for the *Yorktown* cruise, Richard called to tell me that Dad had developed aspiration pneumonia. He had coughed up food from his stomach, and it had entered his lungs. I immediately flew out to California so Richard and I could take turns in the hospital at Dad's side as he slipped away.

He died on March 2, 1998. We had a memorial service for him in mid-March, and I made sure that Mom, who was in good health, and Nancy Ann, who was healthy but always needed special care, were OK before I turned my attention back to *Yorktown*. Luckily, Richard's son, Jeff, lived just a few blocks from Mom and Nancy Ann, and he took care of them for many years in ways that I never could have—visiting every Sunday, bringing them groceries. I'll always be indebted to him and his wife, Carol.

—

WE INVITED FOUR World War II veterans—two Americans who had served on the aircraft carrier and two former members of the Japanese Navy's flying corps—to join us on the *Yorktown* mission. I knew they would be reliving so many emotions. We had found *Titanic* in 12,500 feet of water after searching nearly 100 square nautical miles, and it had taken two expeditions to find *Bismarck*, 15,700 feet down. *Yorktown* lay at nearly 17,000 feet, the deepest I'd ever searched for a ship, and the potential search area spread over 200 square miles. It was perhaps the biggest long shot of any of my searches.

The first decision I had to make was which sonar system to use. One candidate was the *Deep Tow* sonar system that Dr. Spiess and his team at Scripps operated. That would mean dangling several miles of cable to tow the device a few hundred meters above the seafloor. When I ran the numbers, I could see that the water drag against the cable would slow my towing speed to one to two knots, and the device would only be able to scan sections of the seabed about 2,000 feet on each side. Not ideal.

The other option was a sonar system operated by the University of Hawaii. It connected to a shorter tether about 250 to 400 feet below the surface, which would allow for faster search speeds. It also could search a much wider swath on each pass. But the resolution would be lower, and it could only find very large objects. *Yorktown* was 809 feet long, but when you're looking down through more than three miles of water, even big things turn small.

Running more numbers, I convinced myself that if I towed the Hawaii sonar at three to four knots, I could find *Yorktown*. But there was one more caveat: We'd have to be searching parallel to the side of the ship to catch it. We didn't know which way *Yorktown* was pointing, so we'd have to run two sets of lines perpendicular to one another.

It was going to be a crapshoot either way. All I could do was play the odds, so I crossed my fingers and chose the Hawaii device.

Now we needed a ship and an imaging system. Fortunately, the Navy came to my rescue again. My friends at Submarine Development Group One, which had helped us with the Guadalcanal project, had a remotely operated vehicle called *ATV—Advanced Tethered Vehicle*—that could dive to 20,000 feet. Given the publicity that the television special would bring, they were willing to lend it to us along with its support ship, *Laney Chouest,* if we covered some of the major logistical costs, which National Geographic agreed to do.

Barbara was planning to fly to Hawaii to join us on the cruise. We had great caregivers, and she thought she'd try to squeeze it in. But the day before she was to leave, Ben fell in his preschool classroom and split his chin open, forcing her to cancel her plans. *Oh, well,* she said, *that's the way it goes when you have kids.*

While I was in Honolulu, I went to the University of Hawaii to talk to the sonar team. It was an emotional moment for me to be back on the campus where I'd gotten my fresh start after Scripps had rejected me. Then I met up with the National Geographic production team at the U.S.S. *Arizona* Memorial at Pearl Harbor. As we leaned on the rail, looking down at the remains of the ship, we could see drops of oil coming up from below—like tears, some veterans say, from the sailors whose bodies are still encapsulated within the hull.

We all flew to Midway Island the next day to board *Laney Chouest.* Everyone was pumped up, and we started a betting pool on how soon we would find *Yorktown.* But I was worried. I lay awake in my berth the first night, wondering if I'd bitten off too much this time.

Instead of starting on the outer edge of the search box and working our way in, as we had done in other searches, I decided to go straight to the spot where a naval historian on board thought *Yorktown* probably went down. Slowly, as each of the initial passes came up empty, a familiar feeling of panic began to build at the base of my stomach. In my head, I knew that as long as we had calculated the odds carefully, we stood a good chance of success—

theoretically. It was that theoretical part that was causing the uneasy feeling.

It took four days to run the first set of north-south passes and a second set east to west. Early the following morning, I went straight to the lab where Karen Sender, one of the Hawaii scientists, had been processing the data.

"OK, Karen, make my day," I said with all the bluster I could muster.

Karen showed me a small dot on the last sonar line from the night before. Then she pointed to a dot in roughly the same spot on the printout from an earlier pass. She'd also checked the pings from the vehicle's depth finder. They confirmed, she said, that the vehicle had passed over a large object in that very spot.

The data indicated that the object was about 750 feet long and stuck up about 43 feet from the seabed. That was close to the length of *Yorktown,* and almost half its height. I had estimated that the ship smashed into the bottom at 40 miles an hour, and it was possible that she had burrowed well into the soft mud, just like *Titanic*'s bow.

Karen's gut feeling was that it was *Yorktown*. Dave Mindell, the Woods Hole engineer who'd spotted the largest Roman shipwreck when he was with us on *NR-1,* agreed. "Looks awfully good," he said.

Hearing the news, the rest of our team crowded into the control center. Now we just needed to lower the Navy's underwater vehicle and take some images to confirm our hunch.

Piece of cake, I thought, especially because we still had 20 days left on our cruise.

But the Navy vehicle wouldn't work right. On a test-dive near the target, it had only descended to 5,000 feet before its electrical system shorted out. Then, after several false starts, it got down to 15,000, but then, almost as if a bomb went off, one system after another went dead. We hauled it up and could see that two pressure spheres meant to protect the batteries that powered the lights had imploded, creating a shock wave that destroyed the other systems.

It was agonizing.

I tried to keep everyone calm. "I get paid to handle crises," I said to reassure them. Members of the Navy's team and Cathy Offinger, my logistics chief, were on the phone, trying to line up replacement parts, when I came up with another source for the lighting gear. Jim Cameron had used the same system when he filmed *Titanic*. I got a message to him, and he promised to ship us what he had.

We suffered through several more days of difficulties until the Navy vehicle finally made it to the bottom. I didn't have a lot of hope. I had conditioned myself to expect the worst. But then the video camera picked up something that seemed promising: mud balls! I perked right up and started giving orders.

The Navy pilots couldn't figure out why balls of mud would be cause for such excitement. But I'd learned from *Titanic* that large clumps of sediment go flying when a ship craters into the bottom.

Then we picked up, dead ahead, a giant object on our sonar, soon followed by our first images of that mighty warship coming out of the gloom.

"Thar she blows," I called out. "Bingo, bingo, bingo!" Ours were the first human eyes to see her since June 7, 1942—23 days before I was born.

Like *Bismarck*, *Yorktown* was last seen upside down as she disappeared beneath the waves. And like *Bismarck*, we could now see, *Yorktown* had corrected herself on her way to the bottom, coming to rest right side up.

We watched on the monitor as the cameras moved over the port side of the ship, then crossed the flight deck, which still looked usable, and glided toward the island superstructure, the carrier's command center, above the flight deck. Bill Surgi, one of the veterans with us, pointed out the radio room and the bridge. He was choking up at seeing his ship again.

"Too much . . . too much," he said, staring at the screen. "All the people that did their jobs . . . I can see them doing them now." Typical of a charter member of the Greatest Generation, his first thoughts were of the comrades he left behind. Standing near him, I could tell he was also thinking about who had lived and who had died.

As we finished photographing *Yorktown,* it occurred to me that, like the Navy at Midway, we could now wave the victory flag over an expedition that could just as easily have failed.

CHAPTER 12

BLACK SEA QUEST

Most of my heroes are mythical, like Captain Nemo, but there have been plenty of real ones, too. One was Willard Bascom—a brilliant Scripps researcher and something of a maverick. A debonair scientist and adventurer, he lived up to the title of Renaissance man by publishing both a book of poetry and also pioneering research in underwater exploration and deep-sea drilling.

Bascom was handsome, looking more comfortable with a scuba tank on his back than a tie around his neck. He had gotten his start mucking around underwater straight out of high school, during the Great Depression, when he worked on an aqueduct tunnel under the Hudson River that would quench the thirst of his hometown, New York City, by pulling water from reservoirs upstate. He attended the Colorado School of Mines but left early and never received a degree, only to rise to the very heights of the ocean exploration world. Bascom had been part of the famous Capricorn Expedition, a project to map seamounts in the Pacific that was Scripps's first use of scuba gear. He also was a command-ing presence. John Steinbeck once chronicled traveling with

Bascom, describing a crew sitting in rapt attention as he drilled into Earth's crust.

I never met Bascom in person, but when his book *Deep Water, Ancient Ships* came out in 1976, I devoured it. His ideas really started my imagination running. Bascom believed that, compared to the Mediterranean, the Black Sea would prove to be a "treasure vault" of ancient shipwrecks because of its unique lower layer of water. Below 1,000 feet, the Black Sea's water was sulfuric and almost entirely devoid of oxygen, making it inhospitable for marine life. That lifeless layer could prove a godsend to those hunting for ancient shipwrecks, because it would keep out the *Teredinidae,* commonly called shipworms—mollusks that eat waterlogged wood. As I had found in exploring old Roman vessels in the Mediterranean, when wooden ships fall to the ocean floor, those suckers find them, multiply, and eat away until all the wood is gone, leaving nothing but empty mollusk shells behind. I'd seen their work on the decking of *Titanic* as well.

But the chemistry of the Black Sea meant no such life-forms would survive. Without those shipworms, we should be able to find fully intact ships sitting on the bottom. Bascom used the kind of mythic language I loved to describe the treasure waiting to be discovered in those toxic waters: "Somewhere, far out beneath the wine-dark sea of Ulysses, there lies an ancient wood ship," he wrote in the prologue to his book. "It sits upright on the bottom, lightly covered by the sea dust of twenty-five hundred years. The wave-smashed deckhouse and splintered bulwarks tell of the violence of its last struggle with the sea."

As enticing as that sounded, Bascom faced a lingering geopolitical problem—the Cold War. The Black Sea was a hot spot, given Soviet territorial claims, so it was always just on the horizon, the oceanographer's nirvana that Bascom couldn't ever reach. Bascom even convinced Alcoa to build an all-aluminum ship, *Sea-*

probe, designed to raise old wooden ships from the bottom of the Black Sea, but there was no way for him to get there safely. It was that very ship that I was testing in 1977 when the piping system we used to lower our cameras and sonar suddenly collapsed.

But the Berlin Wall had come down in 1989, and the Soviet Union was collapsing. The geopolitical landscape was changing. At the same time that I began thinking more seriously about testing Bascom's theory, two Columbia University professors released a book in 1998 that added another dimension to the lure of the Black Sea. William Ryan and Walter Pitman, authors of *Noah's Flood: The New Scientific Discoveries About the Event That Changed History,* had spent a half decade using core samples to try to prove that the Black Sea region had undergone a massive flood, which they believed to be the historical origin of the tale of Noah's ark. The idea of a massive flood wasn't unique to the Bible, they pointed out. The *Epic of Gilgamesh,* written in the 18th century B.C., also described a flood that wiped out nearly all living things. The Bible even suggested the location for the flood, stating that the ark ultimately rested on the slopes of Mount Ararat, in northern Turkey, less than 200 miles from the shores of the Black Sea.

Ryan and Pitman thought that the Black Sea had in fact once been a freshwater lake, separated from the Mediterranean by a natural dam near present-day Istanbul. About 7,600 years ago, they hypothesized, the dam burst, creating a massive flood that poured salt water into the lake and lifted the water level by more than 500 feet, burying the civilization along the lake's coastline. Some of their core samples seemed to verify an abrupt shift from freshwater to saltwater sea life around that time.

I had a lot of respect for both of them, and I appreciated that Ryan was one of the outsiders who had sat on the committee that awarded me tenure at Woods Hole. He and Dr. Spiess from Scripps had joined up with that crazy Texas oilman, Jack Grimm, to hunt for *Titanic* before I did, but I didn't hold that against him.

The chance to find ancient shipwrecks in the Black Sea was irresistible. I've always loved myths and legends, and this time I truly saw myself sailing in the wake of Jason and his Argonauts, who had sailed the southern shores of the Black Sea in search of a gold-speckled fleece. In the decade since I'd found my first ancient ship in the Mediterranean, I'd only seen Roman vessels. I was itching to find ancient ships from other periods of history. But before I could get to the Black Sea, another ancient civilization came calling.

—

I GOT A CALL FROM CHAS RICHARD, the officer in charge of *NR-1,* who had commanded my recent Skerki Bank expedition. He wanted me to come to the submarine base in Groton and look at video from the eastern Mediterranean. They had been there looking for *Dakar,* an Israeli submarine that had sunk in 1968. I knew the story well—I had used it as part of the plot line for *Bright Shark,* a novel I had written while searching for *Bismarck* years before.

But Chas wanted me to look at the images of mysterious objects they'd seen on the seafloor while searching for *Dakar*. Sure enough, I could make out the shapes of amphoras—those narrow-necked jars, the shipping containers of the ancient world—and they seemed to be lying on the seafloor in patterns that outlined the rough shapes of ships.

There was something interesting there, but I wasn't sure what I was looking at. I took the video up to Harvard and showed it to Larry Stager, an archaeologist who had been excavating the ancient port city of Ashkelon, north of the Gaza Strip.

"Oh, my God, Bob," he said. "These are Phoenician. This could be the most important underwater discovery of our region."

The Phoenicians, based in the region that is now Israel and Lebanon, had ruled the seas for several hundred years, starting around 1200 B.C., long before the Roman Empire flourished. They did not have a giant nation, a great army, great wealth, or vast natural resources, but they did have cunning and a mastery of the sea, which made them a dominant civilization in the Mediterranean at the time. They also developed one of the oldest known alphabets, a variant of Egyptian hieroglyphics, which served as the precursor for many of the languages in written history.

In short, these shipwrecks could prove to be the oldest ever found in deep water. This was what I'd been dreaming about ever since I'd found *Titanic:* going far back in time and finding human history underwater.

Finding such treasures would be a coup for my new Institute for Exploration, and it would be a return to more scholarly discoveries, after my military-focused *Yorktown* mission. As much as I loved the freedom of my new setup in Mystic, I missed the academic life at Woods Hole, the lively debates and occasional jousting with other scholars over what our discoveries meant. In fact, as I looked at those videos, my mind was already racing ahead, thinking about collaborations between archaeologists and oceanographers and wondering if there might not even be a whole new field here.

So I quickly made plans for the ultimate three-fer: I'd spend part of the summer of 1999 checking out what the Navy had found, then I'd head into the Black Sea for a series of expeditions to look for signs of Ryan and Pitman's flood, and I would thereby see if I could prove the Black Sea theory proposed by my hero, Willard Bascom.

—

MY WORK HAS ALWAYS INVOLVED cross-pollination between science, the Navy, and exploration. I feel comfortable crossing those boundaries. On the other hand, I've also found that a lot of people don't like it when I wander into their worlds. Critics complain that I don't play by their rules, that I operate outside the box. That's true. I only have one rule of my own: I don't work with jerks. I find my own way, but I also love to have a guide, someone who knows the world I'm operating in and has the credentials to help open doors. Larry Stager was one of those.

Larry and I had hit it off from the moment I went up to Harvard to show him that fuzzy Navy video. As we shoved off from Malta in June 1999 to find the Phoenician shipwrecks, I was relishing the chance to spend time with him. With an infectious laugh, almost a giggle, and clad in his uniform of khakis, a collared shirt, and an Australian slouch hat—one side of the brim pinned up—Larry was a sight. He'd stare at you behind glasses so thick you'd think they could distort the space-time continuum, and he had a knowledge of ancient civilizations that was hard to match. He also had an uncanny memory for detail.

Larry, then 56, had spent more than 15 years unraveling details about the Phoenicians, a society that had left very little record of its own, even though many ancient writers mentioned it. Larry gave me the royal tour of the massive dig in Ashkelon, where he and his students were painstakingly piecing together the history of how a relatively small society could have held such influence over the Mediterranean.

The Navy had shot the video about 30 miles offshore from where Larry was excavating. Larry's dig had financial backing from two New York philanthropists, Leon Levy and Shelby White, who also were interested in our expedition. Like Anna McCann with the Roman wrecks, Larry could tell me as we searched the site if anything we found mattered enough to recover. The Navy had not given us precise coordinates, just a one-kilometer radius, so we had to use our towed sonar to hunt for the targets.

We had chartered *Northern Horizon,* the same trawler we'd used for the *Lusitania* expedition, but as often happens, we quickly ran into technical problems. A generator died. When one of my team members told me that there wasn't an estimate on when we'd get it back up, I cracked a joke: "OK, got a hand crank?" I got a stone-faced reply. Everyone was tense.

We eventually were able to reroute power, and after some sweeps, our sonar gave us three targets that roughly lined up with what the Navy had told us. As *Jason* dove on the first target, we gathered in the ship's control room. We always keep it dark and cold in there, so we can easily see what's happening on the bank of video monitors and so they don't overheat.

I was squeezing through the tightly packed crowd, moving back and forth between the sensor displays, trying to zero in on the target. It was quiet, nothing but my voice giving directions. Finally, we started to see shapes appear out of the darkness. Larry let out a little laugh, and we gave each other a big hug.

"That's not geology," I said as the man-made outlines of a shipwreck came into view.

We caught sight of an anchor, and then an anchor chain. But the spark of joy immediately went out. The items were clearly from a much newer ship, maybe something from the Victorian era. "We don't care about this guy," I said. "Let's drive as fast as we can to the second target." It wasn't the first time I'd been wandering around the dark ocean floor and come up with something other than what I was after, but it was always a bit deflating.

The sonar signature from the next target was as bright as could be, glowing like a Christmas display in the middle of the screen. *Jason* was showing us a massive stack of amphoras. Applause and whoops broke out in the cramped control room, and Larry just said, "Fantastic!" He took one look and immediately confirmed that the amphoras were from the eighth century B.C. His hunch from the blurry Navy video was borne out.

"It's now your problem, Larry," I said teasingly.

"It's a problem I like," he said. "This is the first Iron Age ship that's ever been found in the Mediterranean."

About 46 feet long, the ship had gone down with roughly 300 storage containers, probably swamped in a storm on its way across the Mediterranean. We could see two cooking pots off to the side, sitting upright as if a meal was on the fire as the ship went down. Shipworms had long ago eaten away the wooden boat, but its stone anchor and clay pots had survived.

It was almost as great a thrill for me to see the smile spreading across Larry's face. "It was ecstasy," he told the National Geographic crew filming the expedition. "I haven't been so happy about an archaeological discovery in years, maybe a lifetime.

"When you have those kinds of moments," he added, "you never forget them, and this was mine."

I knew what Larry meant. It was wonderful when we found *Titanic,* when we found *Bismarck,* when we found *Yorktown.* But in my mind, they really weren't discoveries in the classic sense. They weren't expanding human knowledge. They were what I call "relocations." We were using advanced technologies to find ships swept beneath the waves in our recent past, but we were hunting for ships we already knew existed. But with these ancient wrecks, we were finding ships no one knew were there and uncovering more about societies still full of mystery. With the Roman wrecks and this Phoenician one, we could pull up intact amphoras— objects that often exist only as tiny shards unearthed at dig sites like Larry's. Our discovery was actually providing new information that let us learn more about these ancient mariners, where they came from and where they sailed.

But first, we had to get the artifacts back on land.

Jason was sitting 1,300 feet down. First, we directed it to take some 800 images, so we could build a photomosaic of the wreck. Then it was time to grasp one of the cooking pots with *Jason*'s new robotic hand, an instrument we'd developed since collecting the Roman artifacts in 1997. It looked like a cross between an oven mitt and an erector set.

We thought we'd designed the new hand with fine dexterity, so that it could grip something like this pot gently. We needed to be sure to avoid the fragile handles near the lip of the mouth and delicately maneuver the pot over to the elevator designed to bring artifacts slowly to the surface. But when my pilot tried to close *Jason*'s grip, the pot slipped and fell on its side. The crew gasped. We had to wait for the sediment to settle to see if the pot was damaged. We'd come a long way to figure out what had been on that fuzzy Navy video, and we didn't want to start damaging history now.

The pot was OK, thank goodness, but I'd lost trust in the new mitt. So we pulled *Jason* up to the surface and switched back to the tried-and-true tool we'd used before, sort of a cowcatcher with netted webbing between the fingers. *Jason* went back down to the bottom on another attempt to pick up the cooking pot, this time with success. We picked up the second pot as well and brought it slowly up into the sunlight.

"Now this is archaeology," Larry said, as he watched on the monitor. "Quick and beautiful. That dog can hunt."

We stood together on the deck, staring over the railing, eagerly waiting for the yellow spheres atop the elevator to appear so we could hook them to a crane and bring the haul aboard. We sent the elevator up and down several times, bringing up a total of 16 amphoras. Larry, his own camera at the ready, let out some belly laughs as the treasure trove first visible in that grainy *NR-1* video saw the light of day for the first time in nearly 3,000 years.

A crew of experts on board launched into the process of conserving the recovered artifacts. They placed them into vats of freshwater to remove the salt water that had been pressed into them by the sea. Only then could they begin the long and intricate drying process. With the objects now in the hands of expert conservators, Larry and I went out on deck. It was a gorgeous night, and we sat together in triumph, plastic cups of wine in hand.

The third site identified by sonar turned out to be a nearly identical Phoenician ship. "Bonus," I told Larry. That site yielded an elegant wine decanter, which only confirmed the identity of the sailors and seemed fitting for our celebratory mood.

As the expedition was winding down, we turned up one more bonus, albeit of a different kind, for our hosts in the Israeli government. I told my sonar operators that we'd clearly found a trade route, and they should keep looking for more shipwrecks while I got some sleep. Before long, they woke me up, saying they had lots of hits.

What could that mean? I wondered. The Phoenicians were expert sailors. They couldn't have lost that many ships along the way.

The sonar hits turned out to be gas seeping out, building up deposits shaped like amphoras stacked on top of one another. We ended up pointing the way for Israel to find a significant offshore oil and gas field, here in the waters where ancient sailors had once ruled the sea.

—

TWO WEEKS LATER, I was off to the Black Sea. I'd been blessed with sonar guidance from the Navy along the Israeli coast, but in figuring out where to go in the Black Sea, I had to hire my own scouts. I also needed to find a fishing vessel on which we could poke around.

I started preparing in the mid-1990s, inspired by Bascom's theory that an undersea museum was lying in a layer that would keep wooden ships perfectly preserved. My commonsense notion about ancient trade routes had panned out with the Roman wrecks. These sailors didn't hug the coastline. They sailed straight for their destinations. So I pulled out a map of the Black Sea and looked for a natural harbor, thinking I'd draw straight lines from there to the major ancient cities we knew about. There aren't many protected harbors in northern Turkey, but the town of Sinop sits on a protective bay, making it seem a likely starting point.

I hired archaeology students, including Fredrik Hiebert, a Ph.D. candidate at Harvard, to survey the harbor and land nearby for signs that Sinop was an ancient port. Fred's now National Geographic's archaeologist-in-residence, and we've worked on plenty of projects together since. Even then, nearly 25 years ago, I remember him as being so fixated on finding what he needed that he probably would have walked into the middle of a battlefield just to pick a flower.

Hiebert and the others spent the summer of 1996 combing the beaches around Sinop for evidence that it had served as a trade hub. The beach was strewn with broken shards of amphoras, Fred reported by the end of the season. They also found several kilns where these vessels were being made by the thousands. Clearly, this had once been a major base of operations.

That was an important first clue. I now had an area to target. But I still didn't know much more about the anoxic layer that made the Black Sea unique, and I still had no way to test Ryan and Pitman's idea of a massive Black Sea flood. For that, we needed to go underwater.

I chartered *Guven*, a local fishing boat. Using a simple echo sounder, we created a preliminary map of the depths around Sinop. Then we deployed our side-scan sonar to create a rough picture of the seabed. It quickly detected a sharp leveling off of the ocean floor at a depth of about 500 feet—an ancient shoreline where a beach would have been. Among other coastlines around the world, we find a series of levels, a step pattern, created by six massive floods over the last 22,000 years. Here, there was only one underwater shoreline. This supported Ryan and Pitman's hypothesis that only one flood had come roaring into the Black Sea, converting its freshwater into a saltwater sea connected to the Mediterranean, as it is today. The question was: When did that happen? So we set out to find out.

We started dredging the muck along the old shoreline. One day I was lucky enough to be joined by George Bass, considered the father of marine archaeology. George is almost a decade older than me. An acknowledged expert on shipwrecks in the Mediterranean, he was a big deal long before I ever got into the undersea game. In 1960, he had taken scuba diving lessons to explore a Phoenician shipwreck in 90 feet of water near the Turkish coast, becoming the first person to apply systematic archaeological techniques in an underwater environment.

I first met him in 1973, when I had invited him to speak at a Sea Rovers' diving clinic about what he was doing in the Mediterranean. He taught at Penn State and then later at Texas A&M, where he set up the Institute of Nautical Archaeology. He also created a summer research station in Bodrum, Turkey, after starting to excavate *Uluburun,* a shipwreck from the 14th century B.C. just off the coast. It was one of the oldest shipwrecks yet discovered, and it sat on a rocky slope in about 150 feet of water. George and his scuba diving team worked on that ship for several months every year from 1984 to 1994, making over 20,000 dives to document its condition and cargo. In the long run, George built a home and established a permanent lab in Bodrum. It was beautiful. College research students would come and stay the summer there, and it was quite the scene. In 1988, as the National Geographic Society celebrated its centennial, George and I joined such luminaries as Jane Goodall, John Glenn, Sir Edmund Hillary, and Jacques Cousteau in being named "pioneers of discovery" and receiving the National Geographic Centennial Award at a dinner in D.C.

I had heard that George had some of the same doubts as other traditional archaeologists about my robotic approach, but George had been criticized himself by land archaeologists when he adapted some of their techniques underwater. He understood you cannot be sentimental about the tools you use or how you apply them. He was eager to come out on *Guven* to see what we were finding.

My plan, I told George, was to dredge along the ancient shoreline, and I predicted we'd dredge up flat rocks. Waves smooth and flatten rocks as they roll up and down on the beach, and after the flood occurred, they were covered by mud in the deeper water now located farther offshore.

Sure enough, when the dredge brought up a pile of muck from 500 feet down, I plucked a rock out of the messy goo. It was beautiful, flat and smooth like the ones you skip on water. I tossed it to George. "I sure did come out here on the right day," he said. I think I softened his heart with that little pebble.

We dredged up shells and sent a robot down to capture footage of the ancient shore layer that we now believed had been buried beneath salt water. But when was it flooded? We carbon-dated the shells; they ranged in age from 5,000 to 15,000 years. Then we determined that all the older shells were freshwater species and all the newer were saltwater. It seemed solid confirmation of the theory that the Black Sea changed from a freshwater lake to a saltwater interior sea about 7,500 years ago, just as suspected. The salt water brought in by the flood was heavier than the freshwater it replaced. It fell to the bottom and stagnated over time, losing its oxygen below about 330 feet. Although we couldn't confirm that this event was the same as Noah's Flood, our findings did support Ryan and Pitman's theory that a catastrophic event had created the odd mix of water in the Black Sea.

We were getting closer to an understanding of where we needed to explore, but we weren't there yet. *Guven*'s colorful captain knew that sea well, and he provided the next important clue. He told me that every winter, when he pulled up his fishing nets, they'd be covered in black, smelly sludge—a telltale sign that the nets had dropped down into anoxic territory, the putrid oxygen-deprived waters that Bascom had suggested would protect shipwrecks from shipworms. And he was hitting this layer in much shallower waters than we'd thought possible.

—

THAT TRIP IN 1999 laid the groundwork for the next big expedition the following year. We brought *Northern Horizon* through the Bosporus, into the Black Sea, and set up shop off Sinop to look for signs of civilization along the ancient lake shoreline—anything we could find that might have existed before the big flood.

What did I expect would still be there? *Well,* I thought, *they probably had goats and sheep. And there's a lot of rocks around Sinop.* So we started to look for stone fences.

Two of the new camera systems we'd been developing were finally ready for this season's expedition. *Argus* was my new imaging sled, with bright lights to illuminate the bottom and a color video camera for scouting. *Little Hercules* was our closer. Connected to *Argus* by a 100-foot tether, smaller and more precisely maneuverable, it carried our first high-definition color video cameras.

Sure enough, when we descended on one of the sonar targets, we found stones that seemed deliberately aligned in a rectangular

▲ *Searching for the wreck of the German battleship* Bismarck, *my son Todd points as we watch the images of the Atlantic Ocean bottom relayed by* Argo, *our underwater camera sled.*

▼ *Early in the 1989 search for* Bismarck, Argo *snagged underwater cables—the black lines in this photograph—adding to the stress of the expedition.*

I found my other half in Barbara Earle, whom I married in 1991. We work together, travel together, and understand each other. Here we are in 2019, visiting the Pacific Island of Tanna, with Mount Yasur erupting behind us.

▲ *Pilot Martin Bowen (in red) uses a joystick to maneuver* Jason, *our remotely operated vehicle (ROV), as artist Ken Marschall and I try to determine its position by comparing views on the monitor with a model of* Lusitania.

▼ *Barbara was game for anything, it seemed. She logged sightings through the night during our 1991* Lusitania *expedition.*

▲ *We found the aircraft carrier U.S.S.* Yorktown *some 17,000 feet under-water, its bridge still recognizable 56 years after sinking during World War II.*

▼ *In 2002, with the dynamic duo of* Argus *and* Little Herc, *we inspected a likely target that turned out to be PT-109. Shifting sand on the Pacific Ocean floor had buried all but one torpedo tube, visible here, and a small portion of the boat's aft port deck.*

Once we launched 1.5-ton Jason *off the deck of the mother ship (above), the remotely operated vehicle went to work, shining a light on Mediterranean Sea artifacts dating back thousands of years, including the ancient ship-wreck we dubbed* Isis *(below).*

Our explorations in the Black Sea in 2000 revealed, among many other things, a perfectly preserved ship. The ribs of its hull protruded from the ocean floor.

Returning to the Black Sea in 2011, we discovered human remains from the third century B.C. 101 meters down, in a zone of water with shifting oxygen levels, where fish had removed the flesh but the water had not dissolved the bones.

For years I had dreamed of finding the wreck of Titanic, *the British luxury liner that sank on its maiden voyage in 1912. In 1985, a top secret Navy assignment to explore sunken nuclear subs gave me the opportunity to follow that dream.*

▲ *Photographer Emory Kristof and I proudly hold the National Geographic Society flag aloft as we celebrate the discovery of* Titanic *in September 1985.*

▼ *Adm. Brad Mooney, chief of naval research, in his summer whites, salutes me as he boards R/V Knorr on our triumphant return to the United States after finding* Titanic. *With the press, military, and Woods Hole officials crowding around, it was a zoo.*

We returned to Titanic *in 2004, guided by all we had learned since our 1985 discovery and employing* Hercules *and its high-definition cameras to document the site.*

▲ *Two sizes of shoes, a hand mirror, and a comb evoke a scene. I keep imagining a mother combing her daughter's hair, unaware of their tragic future. Tiles and a bowl tell us these were travelers in a second-class room.*

▼ *Memories of leisure and luxury strew the Titanic site, such as the wrought-iron sides of deck benches, their wooden seats long rotted away.*

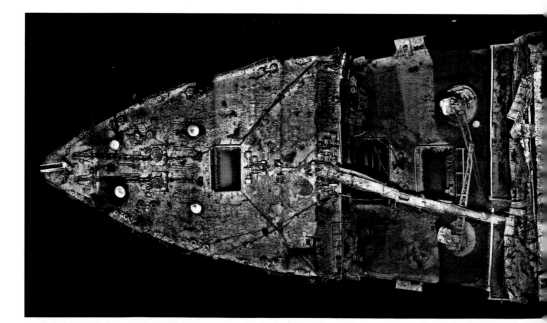

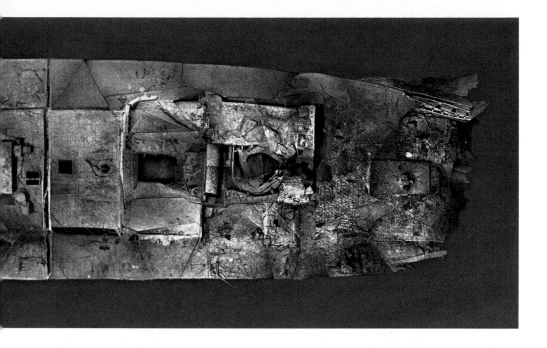

Mosaics of the Titanic *deck in 1985 (above) and 2004 (below) reveal damage over those 19 years. The crow's nest on the mast, near the bow, disappeared. The aft funnel suffered significant damage. Some say time caused the damage, but I blame the numerous submersible dives since 1986, carelessly banging into the ship. Visitors are literally loving* Titanic *to death.*

▲ *In 2019,* Nautilus *plied the Pacific waters off the island of Nikumaroro, searching for any sign of Amelia Earhart's lost plane. In the cool, dark control room, we kept a 24-hour vigil.*

▼ *We launched* Hercules *off the deck of* Nautilus, *sending the ROV in search of any vestige of Earhart's plane.*

▲ *We also deployed an ASV—autonomous surface vehicle—to search close to the reef, where we dared not take* Nautilus. *The telltale site of the* Norwich City *shipwreck is visible on the reef, with a sliver of an island in the background.*

▼ *Drones filming from above reveal* Nautilus *and, below,* Hercules *as it scours the ocean floor day and night for any sign of the famous aviator who disappeared in 1937.*

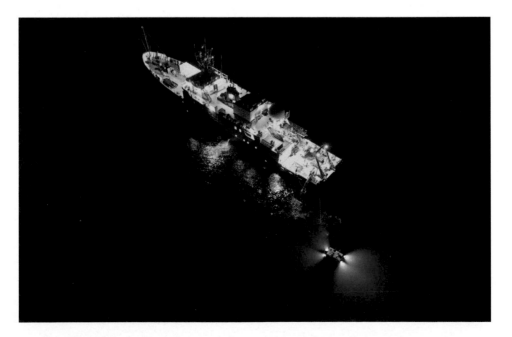

We finally had to leave Nikumaroro behind, disappointed by finding no trace of Amelia Earhart but determined to continue the quest in years to come. At least now we know where she isn't.

shape about 24 feet long by 12 feet wide. It seemed like a pretty good sign of human habitation, possibly even the remains of a Stone Age house. We'd not only found the shoreline, but also a bit of evidence that people had lived along it.

I phoned Bill Ryan to tell him the news that we'd found underwater confirmation of his and Walter Pitman's theory about an ancient flood. He was delighted. Then somehow word leaked out, and headlines began announcing that we'd found Noah's house.

Reporters were hopping into dinghies and rowing out to find me, so many that it started to look like some kind of naval attack. I contemplated whether we needed to pull out the fire hoses to defend ourselves. It was another circus.

I had to get back to the United States, but my crew continued on in search of our dream of a well-preserved ship. I ended up watching a video feed beamed to my command center in Mystic—telepresence in action—as my team surveyed the target we called *Sinop D,* a potential ancient shipwreck about 1,100 feet down.

I've seen millions of miles of mud. I'm used to seeing a crab run across here, track marks there, and little benthic burrows— the hidey-holes of creatures that live on the ocean bottom, under the mud. But in the video from that anoxic layer, there was nothing. The bottom was dead—no critter tracks, no critter holes, nothing. Just the constant snowfall of dead marine life floating down from above.

Then, out of the blizzard, a small forest began to take shape in front of us. At first, we could see only outlines, and it could have been driftwood that had settled down on the bottom. But a pattern began to emerge: Lumber neatly organized in row upon row of evenly spaced handiwork. It took a minute of squinting, but I began to realize we were looking at the outlines of a nearly perfectly preserved ship. There were still knots on the mast from the rigging, and you could see the tool marks from the carpenters who had worked the wood.

Everything was intact, just as Bascom had predicted. With its 35-foot mast upright, the ship looked like it could have been sailing just days before. I was mesmerized. Even when you're a scientist who's done this a zillion times, you are still overwhelmed by every new discovery when it happens.

That clinched it—now I was going back to the Black Sea for sure—and I was going to figure out a new university affiliation. I loved running my own show at the Institute for Exploration, but I longed to reenter the academic world. Working with scholars like Larry Stager and Fred Hiebert had added a richness to exploration that I wanted to repeat. I wanted to bring archaeology and oceanography together and train students in both disciplines in a single program.

I thought about my longtime hero. For so long, my dream had been to find the perfect ship in the Black Sea, and then call Willard Bascom to tell him about it. When we found *Sinop D,* I asked my team to find his number.

They came back with a tragic message. Willard Bascom had just died, complications after a car accident. He died within days of my being able to tell him that the idea he had proposed decades before, the theory that had inspired me to pursue his vision, had been right.

CHAPTER 13

CLOSE-UPS ON THE PAST

B y 2001, Barbara and I had a solid history of success with the specials we were producing for National Geographic Television and Turner Broadcasting. They were constantly pressing us for new ideas, the more ambitious the better, because they wanted at least one special each year. For 2002, we decided to focus on one of the most iconic shipwrecks out there: Sir Ernest Shackleton's *Endurance.*

Roald Amundsen had made it to the South Pole in 1911. After that, Shackleton decided to lead an expedition across Antarctica, coast to coast. Shackleton and his 28-man crew set out on the Trans-Antarctic expedition in 1914, a few months before the onset of the First World War. Their three-masted ship, *Endurance,* got stuck in the sea ice before they could even reach the continent, much less cross it.

A humiliating failure on its face, it became one of the epic tales of rescue at sea. After being trapped on the ship for months in the drifting ice pack, Shackleton and his crew abandoned it right before it sank and took off on a torturous, multiday journey to a nearby island, part dragging and part sailing the support boats through the breaking ice floe.

From there, Shackleton took a skeleton crew on a nearly impossible, 800-mile, open-boat trip to remote South Georgia Island, finding their way to this tiny speck in a vast, stormy ocean. There, he organized a rescue party that went back and retrieved the rest of his crew, all without losing a single man.

Talk about an iconic target. Shackleton and his men left *Endurance* behind as the sea ice crushed, then engulfed it. Photographs taken in 1915 show the massive sailing ship crumbling under. No one had ever located the wreck. It's a target of immense historic interest, no doubt about that, and any expedition to find it would require weeks in some of the wildest and most spectacular seas on Earth.

But it wasn't to be.

We were getting things all lined up, but when a friend at NASA told me that forecasters predicted especially severe conditions that summer, I pulled the plug. It was disappointing, and we still had to come up with a television special for 2002.

Barbara and I scanned all possibilities. I wanted to go back to *Titanic* to take higher-resolution pictures and gain closure, using our powerful new camera system, *Hercules*. But *Herc* wasn't ready yet. I had funding to explore Lake Huron's Thunder Bay. It's got so many sunken ships that it's nicknamed Shipwreck Alley—but National Geographic wasn't keen on that idea. The Black Sea? We were already set to return there in 2003, and National Geographic was happy to wait. Inca treasure lost underwater in Peru? Again, the television people weren't sure. What about the 1845 Arctic expedition led by Sir John Franklin, lost in ice-choked Canadian waters? Or U.S.S. *Indianapolis,* on which nearly 900 sailors died in the shark-infested waters of the Philippine Sea when a Japanese submarine torpedoed the ship after a top secret mission to deliver the atomic bomb?

Then somebody mentioned *PT-109.*

The patrol torpedo boat known as *PT-109* had been lost on August 2, 1943. It inspired a legendary tale of courage at sea during World War II, and became a centerpiece in the political career of its young skipper, John F. Kennedy. But the 80-foot boat had never been found. Locating it would be of immense historical interest— but it would not be easy, a true needle in a haystack. Not only was the boat less than one-tenth the length of *Titanic* or *Yorktown,* but there was also a bit of a mystery about what kind of damage it suffered and where it sank. Perhaps if we found it, we could resolve those questions.

The story had special resonance for me, having lived for so long in Cape Cod, near the Kennedys' Hyannis Port compound in Massachusetts. I'd gotten to know Ted Kennedy and other family members, and my son Ben was born at the hospital in Hyannis. There were political ramifications as well. *PT-109* was legendary, and finding it would boost my efforts to lobby Congress for more funds for ocean exploration.

I was in Washington, D.C., when we made the call to go with *PT-109,* and I happened to meet the next day with Fritz Hollings, a Democratic senator from South Carolina and a pivotal figure in funding ocean research. Hollings had sponsored the congressional earmarks that had funded *Hercules,* as well as other systems I was developing to replace the ones I had left behind at Woods Hole. He had grown close to Jacqueline Kennedy after her husband's assassination, and he was excited to hear about the *109* search we were planning. Jackie had given him some of JFK's personal effects, he said, but he had lost them all in a house fire.

It wasn't the first time I'd thought about searching for Kennedy's boat. During our work in Iron Bottom Sound, in Guadalcanal,

I had hoped to make a side trip to look for *109,* but we ran out of time. That meant we had done some research already, and we wouldn't be starting from scratch.

Still, I remained wary. Exploring the Black Sea was my main interest at the time. *PT-109* felt like a sideshow. Did we even have enough time? I checked my calendar for 2002. It was ugly. I had multiple obligations, including serving on President George W. Bush's Commission on Ocean Policy. The way it looked, I could only spare a couple of weeks for the *109* search.

But Turner and National Geographic needed another Bob Ballard special, so *The Search for Kennedy's* PT-109 was on.

And I found myself exactly where I did not want to be, floating again in the sweltering archipelago of the Solomon Islands. It had all come back to me as the plane dropped out of the clouds and the dense rainforest of Guadalcanal came into view. I flashed back to all those technical nightmares—busted cable, running aground, overheated generator, black smoke pouring out of the engine compartment. Yet here I was in a place to which I'd sworn never to return.

We contacted a local American resident named Danny Kennedy—no relation to the president—who had done some research work for us during our earlier Guadalcanal expedition. He was eager to join the search. Cathy Offinger, my logistics chief, found a ship, *Grayscout,* for rent in Australia. National Geographic came up with the $24,000 we needed to reserve the ship, and we started spending money and hiring people at a rapid clip.

Just as our efforts hit high gear, John Fahey, then president of the National Geographic Society, voiced concerns about the Kennedy family. Would they be upset? He had to find out, and if the Kennedys objected, the project would be canceled. Not great news for us as money was flying out the door.

Most of John's discussions were with Caroline Kennedy, JFK's daughter, and his surviving brother Ted, both of whom Fahey knew. They were concerned that we might disturb the site or attempt to recover objects. We assured them that we had no intention of disturbing *PT-109,* only locating it. For one thing, the Navy regarded the sunken craft as a grave site. For another, we weren't scavengers; we were explorers, seeking knowledge. Finally, the family agreed—as long as Max Kennedy, the ninth of Robert's 11 children, could come along with us.

—

THE SOLOMONS—SIX MAJOR ISLANDS and nearly a thousand smaller ones, some just a few palm trees clinging to coral—begin east of Papua New Guinea and swing more than 900 miles to the southeast in two parallel volcanic chains, with a narrow stretch of water between them known as the "Slot." During World War II, the whole archipelago was like a ladder leading from Allied-controlled Australia into the heart of Southeast Asia. Island by island, America and its allies picked their bloody way north until, by the summer of 1943, the fighting concentrated in the northern Solomons, where the target was the Japanese supply route.

Because the Allies controlled the air, the Tokyo Express—the American nickname for both the supply route and the ship that ran it—made its irregular runs at night, slipping down the Slot with an escort of Japanese destroyers, unloading its desperately needed cargo of men, weapons, and supplies, and then hustling back up the Slot before dawn. The PT boats (short for patrol torpedo) were designed for close combat with larger ships, and they regularly tried to intercept those night runs. And Japanese destroyers, when they saw PT boats, regularly attempted to run them over.

On the night of August 1, 1943, *109* was one of 15 PT boats in the Slot. Aboard were Kennedy (its captain), two junior officers, and 10 other crew members. In the middle of the night, sure enough, the Tokyo Express and its escort came sliding out of the fog. The

PT armada attacked but missed with every shot. The Americans regrouped. The boats that had fired all their torpedoes headed back to base, leaving *PT-109* and two others behind. The three lined up, creating as much of a barrier as they could on such a foggy night.

Finally, at 2:30 a.m., Kennedy spotted what he thought was another PT boat emerging from the murk. It was actually *Amagiri*, one of the Japanese destroyer escorts, and by the time he realized his error, it was too late. Kennedy tried to swing *109* around to face *Amagiri*, positioning the boat so he could fire his torpedoes, but the destroyer was too fast and slammed into the PT boat, which exploded in a ball of fire. Two crewmen died instantly.

Now the fog-shrouded sea was flecked with pools of burning fuel. A portion of the stern had been sheared off in the collision and sank. The surviving crew members clung to the remains of the boat as Kennedy swam repeatedly into the flames to rescue others, including one with burns over 70 percent of his body. They hung on the wreckage for 12 hours, drifting slowly south and perilously near the Japanese stronghold of Kolombangara Island. The following day, with the hull close to sinking and Kennedy fearing the Japanese would spot them, he led a risky swim across miles of Japanese-patrolled waters to a tiny, uninhabited island. Kennedy kept the wounded man afloat, clenching the strap of his life jacket in his teeth as he swam.

Days later, the 11 men were discovered by two friendly local fishermen. Without a radio to signal for help, Kennedy improvised, carving a message into a coconut shell: "NAURO ISL . . . COMMANDER . . . NATIVE KNOWS POS'IT . . . HE CAN PILOT . . . 11 ALIVE . . . NEED SMALL BOAT . . . KENNEDY." The local men took the coconut—which was the way word eventually got to Navy troops based on Rendova, another island. They heeded the call and rescued the crew of *PT-109*.

—

I LANDED AT GIZO on May 17 in a raging downpour, typical for that part of the world. *Grayscout* was there, so by late afternoon we had loaded our equipment and were on our way out to look for *PT-109*.

We had a pretty good idea of where the collision took place, based on accounts from other PT boat skippers and Reginald Evans, an Australian in hiding on Kolombangara, assigned to monitor Japanese shipping activity. The collision happened in Blackett Strait, which feeds into the Slot off the island's southern coast, and *109* was the only PT boat that had been lost in that area.

Early reports indicated that *Amagiri* had sliced *PT-109* completely in half. That's what Kennedy thought, and that's the story that had captivated the public when he ran for president. That would mean the boat's stern had sunk near the collision site, while the bow had floated south with the current, the crew clinging on. Indeed, the bow section of a wrecked ship had been spotted on a coral reef not long after the event and had long since washed away.

Evans believed a PT boat had attacked a Japanese barge and caused the explosion. When he learned a PT boat had been hit instead, he enlisted local fishermen to help search for the wreckage and any survivors. He reported having seen a floating object south of the crash site in Blackett Strait the next day, but it was too far away for him to see if any crew were clinging to it.

So we didn't even know quite what we were looking for. Some historians speculated that the destroyer had only sheared off the starboard side of the boat, rather than knifing through the center, because Kennedy reported he was in the middle of the boat when the collision occurred. That would mean that an even smaller— and harder-to-find—piece remained on the ocean floor of the collision site.

Either way, what we knew for certain was where the crash had occurred. The wisest plan was to start by looking right there.

From *Grayscout,* we began a systematic run of parallel sonar lines down the axis of the strait. We were towing our sensor in water over a thousand feet deep, so we had to move slowly to keep the equipment from kiting. We kept up round-the-clock operations, and the change that overcame Blackett Strait at night was striking. The scattered fires from remote, tiny villages flickered out, and the lush islands disappeared into a void, as though you'd entered a pitch-dark box with no signs of life.

At first, I thought the wreckage of *109* might be large enough to stand out, but as time passed, that all-too-familiar terrible feeling settled in the pit of my stomach. We were finding hundreds and hundreds of objects on a seafloor more complex than I'd anticipated. Steep coral walls tower hundreds of feet above the seafloor, and countless blocks of coral litter the bottom, with endless fields of man-made debris scattered everywhere. How were we going to visually inspect all these targets? There simply was not enough time.

Through Dan Kennedy, we met Eroni Kumana, one of the two fishermen who had found the stranded crew. When we introduced him to Max Kennedy, the old man broke down in tears, overcome by meeting someone he called a "real Kennedy" so many decades after the crash. We traveled with him into the jungle to meet Biuku Gasa, the other fisherman with him that day. Eroni then took us to Nauru, the remote island where he and Biuku had first found Kennedy, who'd swum there in hopes of flagging a passing American boat.

It was a tiny outcropping, and Eroni went straight to the spot where he found Kennedy. He pointed to a tall palm, saying excitedly that this was the very one from which Kennedy had taken the coconut to carve his rescue message. They said they had explored the wreckage on the coral reef that many thought was *PT-109*'s bow, and they had found something odd in it: Japanese rifles.

That made me rethink the whole scenario. What if *109* had not been cut in half? What if, as some researchers speculated, *Amagiri* had only clipped off a back corner of *109*? That would leave a much larger section of the boat fatally stricken, but floating. If the wreckage on the coral reef was not *109,* but a Japanese ship instead, as Eroni and Biuku thought, then *109*'s bow could still be out there somewhere.

Over the years, I have learned to trust what local witnesses say and not overthink these things. I went back to the report from Reginald Evans and plotted the location of the object he reported, using his rough estimate of its location out at sea. Then, to create a search box, I calculated how far that object could have drifted. We moved *Grayscout* south and established a new sonar search grid. The ocean bottom here was different—smooth and empty—and our sonar found just one long, lonely target in 1,300 feet of water.

We launched *Argus* and *Little Herc,* our pair of imaging robots. *Argus* hangs directly below the ship, equipped with powerful lights. *Little Herc,* connected by a tether to *Argus,* explores the ocean floor beneath it. It's like *Argus* is walking a dog on a leash, and it's important not to tighten the leash abruptly and pull *Little Herc* off the ocean floor.

As we approached our lone target, the deep-sea sonars quickly picked up signs of wreckage. The sonar can penetrate several feet into the sandy bottom, and it revealed an object measuring 40 feet by 23 feet. *PT-109* was 23 feet wide, but 80 feet long. Perhaps the rest of the ship was buried too deep for the sonar to reach? *Little Herc*'s cameras pulled into range, showing us a long metal cylinder the shape of a giant cigar on the seafloor.

Fortunately, we had two key people with us: Dale Ridder, an expert on PT boats, and Dick Keresey, who had been the skipper of *PT-105,* also positioned on Blackett Strait when *109* was hit. They were convinced the cigar-shaped object was a torpedo tube with an unfired torpedo still in it. The torpedo's propellers were visible. They looked like the kind on PT boats. Then I heard what I was waiting for when Dale said, "That's *PT-109!*"

Dale noticed a mounting strap on the torpedo launcher. It appeared to be attached to a wooden block. Maybe the deck was there too, but buried in the sand. PT boats were made largely of wood, and many researchers had fretted that any remains of *109* could have been eaten away. That didn't worry me. I knew that the deck was mahogany, which is not on the menu of the wood-boring mollusks that ate the yellow pine deck of *Titanic.*

We asked Max to speak for his family and give us permission to shift some sand near the torpedo tube to see if we could find a buried piece of deck. With their approval, we rigged a small plow on the front of *Little Herc* and used the robot, slowly and awkwardly, to push away the sediment. There was the torpedo's air flask as well as the opening in the deck for the torpedo gyroscope.

We continued to explore gently. Our sonar showed other large objects nearby, and we were finding a lot of metal. Our makeshift plow uncovered a cranking mechanism used to launch the torpedo. We tried to push it forward. No movement. We tried again. It refused to budge. That meant the mechanism was still attached to the deck. Blackett Strait has a strong current running through it, and it appears that after *109* landed on the bottom, waves of sand moving across the bottom mostly covered it, much like sand dunes bury a fence along a shoreline.

We did not disturb the site any further. The final resting place of *PT-109* was no longer a mystery. Mission accomplished.

I was not unhappy to be leaving the Solomons, and I had some nice souvenirs to bring back. One evening, several canoes had approached *Grayscout,* local artisans from whom I bought some fine stone carvings. I also asked one of the boys if I could buy his hand-carved canoe. After consulting his mother, he agreed. I gave him $100. He handed me the paddle and dove into the water. We wrestled the canoe on board and put it in one of our shipping containers. Unfortunately, customs officials refused to let me bring it into Australia, fearing destructive insects were concealed in the wood. But I still have my hundred-dollar paddle.

It's not the most precious souvenir I brought back, though. When Eroni took us to Nauru Island, I had picked up two coconuts at the base of the tree where he had found Kennedy. Back in Mystic, I hired a wood carver to make replicas of Kennedy's hand-carved message on them. One I still have. The other I presented to Fritz Hollings, whom I invited to our private screening of National Geographic's television special, *The Search for Kennedy's PT-109.* I wanted to help him make up for what he had lost in the fire, and he was moved by the gesture.

—

AS I STEPPED UP my lobbying efforts, seeking more funds for ocean exploration, I needed to build strong personal connections with lawmakers like Fritz Hollings, and I had to stay in the public eye with news-making expeditions.

Hollings had played a major role in creating NOAA—the National Oceanic and Atmospheric Administration—when Richard Nixon was president. He loved Jacques Cousteau, and he kept calling me "the new Cousteau." The fact is, I explored much deeper parts of the ocean than Cousteau—the parts that were less interesting to most people unless they contained an important piece of human history. But I appreciated that Fritz was smart enough to see that our field was contributing just as much to the public's understanding of the oceans as Cousteau had. He and

Senator Judd Gregg, a Republican from New Hampshire, were the leaders of the Senate Appropriations Subcommittee that oversaw NOAA's budget, and both supported what I was trying to do. We were going to need a lot more help from other players in Washington, D.C., however, if we were going to have the kind of ocean exploration program that I thought the wealthiest nation on Earth should have.

I had served on a panel that had drafted a national ocean exploration strategy for President Bill Clinton in 2000, and we had recommended the creation of the Office of Ocean Exploration within NOAA, part of the Commerce Department better known for running the National Weather Service. We had suggested funding the program at $75 million a year for 10 years. But in its first year, Gregg and Hollings were able to secure only four million dollars. I kept campaigning, and now, as I returned from the *PT-109* search, I was serving on a similar commission created by President George W. Bush, with two more years of meetings and public hearings. I loved the experience—talking to experts in so many other ocean-related fields. It was like going back to graduate school.

At the same time, I was rejoining university life. I needed an academic home for the explorations I was mounting from my institute in Mystic, and I missed the camaraderie of other scientists I'd had at Woods Hole. I worked out a deal with Robert Carothers, president of the University of Rhode Island, the school where I had earned my Ph.D. I received a tenured research position as professor of oceanography and a $100,000 annual budget to create a new Institute for Archaeological Oceanography. I'd never thought of oceanography as a separate discipline like physics, chemistry, or geology, but rather as a field in which all these disciplines intermingled. With my searches for ancient shipwrecks, I'd informally added archaeology to the mix. Now I could formalize that connection and realize my dream of creating a new academic discipline—one in which students could receive a master's in marine archaeology and a doctorate in oceanography at the University of Rhode Island.

My worlds were expanding once again. *Hercules* was finally ready for action. The most sophisticated vehicle we'd developed at my institute in Mystic, *Hercules* was the first robot designed specifically for remote undersea archaeological excavations. Bright yellow and weighing in at 5,200 pounds, *Hercules* had high-definition video cameras, rugged manipulator arms that could lift heavy artifacts without scratching them, and a suction system for excavating ancient shipwrecks according to archaeological standards.

All these working parts were coming together in 2003, as I planned another expedition to the Black Sea and the Eastern Mediterranean. Key people from my institutes in Connecticut and Rhode Island would be working together to test *Hercules* and an advanced ship-to-shore communications system, so scientists and the public could interact with us as explorations unfolded. We had government and private funding, National Geographic was planning TV shows and a magazine story, and the JASON Project would produce online educational programs.

But my luck went south. First the Turks held up research visas for my crew, which kept us tied up at the dock in Sinop, wasting $40,000 a day in operating expenses. We ran some initial tests with *Hercules,* mapping a couple of ancient shipwrecks and retrieving mud samples from *Sinop D,* the perfectly preserved ship from a thousand years ago that I'd hoped to tell Bascom about before he died. But we had to move on to keep to our schedule— only to find out that Egypt was refusing to give us permits to excavate the two Phoenician ships we had discovered several years before in the Eastern Mediterranean. It was frustrating, and we didn't get much work done that trip.

Back home, I kept on lobbying, pressing the government to up the ante on ocean exploration. Chris Dodd, a Democratic senator from Connecticut, brought his daughter for a visit to Mystic Aquarium in January 2004. Over dinner, the senator spoke of his admiration for Senator Hollings, who had just announced that he would not seek reelection due to his wife's failing health.

"One of my goals in life is to try to fill one of Senator Hollings's shoes," Dodd said. "How can I help you?" He invited me to his house for dinner the next week, and there I explained that, in addition to providing exploration grants to scientists, NOAA needed to have its own dedicated ships of exploration to extend how long researchers could conduct their work at sea.

Senator Dodd suggested I meet one of his colleagues, Daniel Inouye, the Democratic senator from Hawaii. We arranged a meeting, and I explained that both the Clinton and Bush ocean policy panels had endorsed the need for dedicated ships of exploration. "Well, NOAA doesn't have a lot of money," said Inouye. "I think our best bet is to get it from the Department of Defense, since they own a lot of ships and have all the money."

Senator Inouye was a close friend of Senator Ted Stevens, the Alaska Republican who chaired the Senate Appropriations Committee at the time. Thanks to Inouye's help, a month later I found myself in Stevens's office. It was truly amazing, not only for its size and grandeur but also for all the artifacts from Alaska. One of his favorites was the penis bone of a walrus, which was the size of a baseball bat. I was told he sometimes used it as a gavel to get everyone's attention.

"The best approach is through the Department of Defense," Stevens said, echoing what Inouye had said about how to provide NOAA with a dedicated research ship. "How much do you need to fix one up?"

"I have been told $20 million should do it," I replied.

"If I get you the ship and $20 million, I want you to use it to find a lot of fish no one knows about off Alaska," Stevens said, probably only half-joking.

With Senator Stevens in my corner, things started happening. The Pentagon decided the U.S.N.S. *Capable,* a general surveillance ship, was the vessel best suited to become a research vessel. It had been commissioned in 1989 to track down Soviet submarines and had later been used as a counter-narcotics ship. The Navy withdrew *Capable* from military service in September 2004 and transferred it to NOAA the same day.

It was going to take several years to convert the ship into an oceanographic research vessel. We christened her *Okeanos Explorer,* named for an ancient Greek Titan, primeval god of the waters. To persuade Congress to put up more money, I got to know Representative Frank Wolf, chair of the House Appropriations Subcommittee that handled NOAA's budget. He was a Republican from Northern Virginia, and Rachel Carson Middle School was in his district. It all seemed fitting. The school was named for the writer who had published *The Sea Around Us,* a best seller about oceans, and teachers there had been bringing students to our JASON Project shows for a decade.

I called the school's principal and said I'd like to visit his teachers and students. "And I would like," I added, "to ask your congressman, Frank Wolf, to introduce me."

Wolf agreed, and he and his wife came to the presentation. The room was packed as I talked about how the JASON Project had just taken students to the rainforests in Peru. I also talked about America's new ocean exploration program. As I spoke, I never took my eyes off the congressman, hooking him, too.

The next time I was on the Hill, I stopped by Wolf's office and reminded him that we needed more money from the House. "You know, Dr. Ballard, you are the first person to ever come into my office and explain the importance of ocean exploration," he replied. "But I can guarantee that after you leave, there will be a line of lobbyists around this building from aerospace companies, waiting to come into my office to tell me how important it is for me to increase funding for space exploration."

It was a reality I knew all too well. The race to put a man on the moon had electrified the country in the 1960s, and space travel had retained that allure. Meanwhile, even though scientists and explorers had been doing equally remarkable feats in the ocean through all those years, we were still struggling to get the word out. We needed the world to understand that as important as exploring outer space was, exploring inner space was equally essential. Fortunately for me, when Representative Wolf retired, he was succeeded as chairman of the House Appropriations Subcommittee by Representative John Culberson from Houston, Texas. Culberson increased funding for exploration to its highest levels and went to sea with me every year to ensure that taxpayers were getting their money's worth.

—

SLOWLY, AS MORE RELIABLE government funding and academic support began to fall into place, it was time to shape my next expedition. I knew that returning to *Titanic* could grab the public's attention, and so I steamed back to the site for the first time in 18 years. *Hercules* and its high-definition cameras were ready to go, so we could take much richer and sharper color images than we could in the 1980s. I wanted to see how much damage the sea—and high-tech scavenger hunters—had done to the grand old lady, and I wanted to say my final goodbye.

After I found *Titanic* in 1985, I could have obtained salvage rights if I'd wanted to, according to international maritime law. But I wasn't interested in retrieving artifacts. I believed we should respect the passengers and crew who had died there and leave their grave site undisturbed. Survivors, like Eva Hart, agreed. She was seven when the ship sank. "I saw all the horror of its sinking," she told me. "And I heard, even more dreadful, the cries of drowning people."

It didn't take long, though, once the location was publicized, for *Titanic* mania to take over. Investors saw dollar signs and formed a partnership, Titanic Ventures, which became a company, RMS Titanic Inc. Then they and others proceeded to plunder the wreck. The French government, still smarting from my discovery and the feuds that resulted from it, wanted to return to the site, so they leased *Nautile,* their deep-diving mini-sub, to the investor group in 1987 and 1993. By 2000, RMS Titanic Inc. had returned to the site four more times, using French or Russian submersibles. In a game of Finders Keepers, they pocketed more than 6,000 artifacts and displayed them in a museum, charging people to see them. The company even broadcast a documentary showing how it took the objects.

All told, the items included eyeglasses, shoes, handbags, luggage, and even a bronze cherub statue from the Grand Staircase. A bell and a light from the foremast were removed, and the salvagers even raised a chunk of the hull weighing 18 tons. They sold pieces of coal from the engine room for $25 a block. They created a website, so you could peruse the collections online. Documentary filmmakers and wealthy sightseers visited the site in mini-subs. And, perhaps most grotesque of all, a couple were married in a submersible perched on *Titanic*'s bow. I wouldn't think of a mass grave as romantic, but I guess some couples are into that.

Without realizing it, we had opened all this up when we'd found the wreck, and it had turned into an ugly carnival, an affront to the fate of *Titanic* and all those who lost their lives in her final hours. It was the exact opposite of how we, the Kennedys, and the

U.S. Navy had treated the resting spot of *PT-109*, not disturbing anything. But *Titanic* was not similarly protected; there was no guard on duty.

I spoke out against the grave robbing and called for international legal protections. An agency within the United Nations approved a convention in 2001, calling on nations to protect "all traces of human existence having a cultural, historical or archaeological character" that had been underwater for more than 100 years. But it had been just 92 years since *Titanic* had gone down, so those protections did not apply.

Against this backdrop of reckless salvaging, I thought that vivid images of the damage might counter the disrespectful gimmicks and rally support for preserving the site. I wanted people to start thinking about the oceans themselves as a museum to be visited and respected, but not pillaged.

So off we went aboard NOAA's flagship, *Ronald H. Brown*, bound for the *Titanic* site once again. We spent several days remapping the area. Then we lowered our robots into the blackness 12,500 feet down, with *Argus*'s bright lights shining on everything around it and *Hercules* tethered to *Argus* to take a closer look.

It was tense enough to be back there, watching *Titanic* come into view again. But what really scared the heck out of me was that National Geographic was insisting on broadcasting one of our dives live. We were still working out the kinks in *Hercules*' technology, and to say "I'm going to come live from the deck of *Titanic*" was a bit terrifying. I was going to leave that dive to the end.

Sure enough, on the first try, we had trouble controlling *Hercules* on the way down, but we quickly stabilized it just before a large object loomed in front of us. Suddenly, I realized it was the rear section of *Titanic*'s bow, right where the great ship had split in two as it sank. I had never visited this part of the ship before, because it would have been very difficult for *Alvin* to go into such a dangerous place. But *Hercules* was much smaller, and with *Argus* giving us a bird's-eye view from above, I could see what was coming.

"That's a boiler!" I exclaimed, echoing what one of my colleagues had uttered when we had first come upon the ship in 1985. But this one was in place, still connected to the ship. Much of the metal had a blue-green patina like you see on old copper pipes. Above the boilers, we saw what was left of the shredded and sagging promenade deck, where the lifeboats had been launched. We moved along and saw the rusted telemotor that had held the wheel a crew member had frantically spun to try to avoid the iceberg. The handrails on the deck had fallen away, either from natural deterioration or human destruction. With our new technology, we were just inches away, and all of the images were sharp and vivid.

On our second dive, *Hercules* went dead at the bottom, thanks to a kink in the line tethering it to *Argus* and its electrical power. So we had to recover the robots and make some repairs—a whole day lost.

The third time's the charm, we said, trying to stay optimistic. And it was true—the third dive was amazing. *Hercules* roamed over the bow section for 12 hours, systematically photographing everything. We also dove on the stern section and in the broader debris field, where most of the personal items had fallen. Objects still scattered the ocean floor: a silver dish covered in a deep green patina, wine bottles, a teapot. And a pair of women's shoes—the two of them, lying right next to each other, both pointed in the same direction. Several combs, a hand mirror, and a child's shoe lay on the ocean floor nearby. It's pretty obvious that bodies had landed there.

We also saw the destruction caused by scavengers. There was the toppled mast, but now it was missing its bell and brass light. A hole in the deck suggested that someone tried to rip out the telemotor. Dents showed where submersibles had landed on the wreck, and tracks in parts of the debris field made it look strip-mined.

It was heartbreaking. I lingered, knowing this might be the last time I would ever spend with *Titanic*. The cameras on *Hercules* and *Argus* gave me both a broader view and sharper close-ups of everything, from scrapes in the hull to growing fields of rusticles. I saw the lifeboat cranes and thought again about how many others could have been saved.

My farewell had a bitter edge, because part of my job was to carry evidence of the devastation back to the world. My first step in doing that came when we did go live, without hitch, from *Titanic*'s deck on the National Geographic Channel. I also spoke to news outlets after the trip, making a plea for the vultures to let *Titanic* rest untouched. "You don't go to Stonehenge and push the stones over," I said.

Her earrings, watch, and brooch are missing, but the old lady is still there. If I had my way, the *Titanic* site would be an underwater museum, and we would use telepresence to stream images from the ocean bottom where she lies. Then it wouldn't just be the wealthy few who get to see her, but everyone in the world could experience *Titanic*.

CHAPTER 14

BECOMING CAPTAIN NEMO

F or all my career, I'd look in the mirror in the morning and say
to myself, *Never own a ship.* It's a hole in the ocean you pour
money into.

Ever since my first cruise as a high school intern at Scripps,
I'd always sailed on ships owned by research institutions, the
National Science Foundation, the private sector, or the Navy. But
after leaving Woods Hole, I had to compete with oil companies to
charter vessels. They had all the money in the world, and they
leased them for six months at a time. As a result, I often ended up
with the short end of the deal—ships I didn't really want, available
just as bad weather was setting in at a starting location far from
my destination. To me, ships were like buses or taxicabs—vehicles
to get me where I needed to go for a month or so, but not the core
of what I did. Maintenance, insurance, repairs, and a crew—those
were someone else's problems.

My toys were mobile. Undersea vehicles, a launch-and-
recovery crane, a satellite system, tools, and a control center:
Everything fit into six cargo containers, and we could install them

on any ship with dynamic positioning capabilities and an aft deck large enough to hold them. We had the system down. A couple of flatbed trucks would carry our containers to New York or Boston and place them on a cargo ship, which would drop them off at a port where our charter ship was waiting. Once the containers were aboard, my team would spend two to three days setting everything up. The crane and tethered camera vehicles would come out of the containers. The command center was built into two vans. We just locked them to the deck of whatever ship we were on, and it was ready for action.

But there were limits to what I could do on a ship I didn't fully control. By 2008, I had made five expeditions to the Black Sea, each on a different ship. I wanted to go back again. Could I find an even older shipwreck, and might the bodies of its crew be perfectly preserved, given the oxygen-depleted water? I wanted to explore new places, too, like the Indian Ocean. I wanted to have the freedom to stay at sea longer than any given charter might allow.

I also needed to reinvigorate my efforts to excite kids about science and engineering. The JASON Project had largely run its course. Now that we had the internet, there was no longer any point in busing students to museums and science centers once a year to watch broadcasts. Why not create something where they could see what our cameras were seeing, live as our explorations unfolded, right on their own computers?

What I really needed was my own ship of exploration. I needed to become Captain Nemo, with my very own *Nautilus,* or at least a new version of Jacques Cousteau and his ship, *Calypso.* But finding a ship I could afford, I realized, might be as challenging as finding ancient shipwrecks.

Several of my wealthy friends had made donations to my sea-going programs, and by 2008, I had built up a war chest of about

$1.5 million. But every time I submitted a bid on a ship that might work, I lost out to an oil company. Then one day in June 2008, Laurie Bradt, who handled the finances for my Institute for Exploration, came into my office in Mystic with something on her mind.

"Bob," she said, "I found a ship for sale on the web that looks like what you want."

It was originally an East German hydrographic vessel—a ship designed to map the ocean—and the current owner, a private citizen, wanted to sell it. Named *Alexander von Humboldt*, after the famous German explorer and geographer, it measured more than 200 feet long and had berths for more than 30 people. It all sounded good to me.

I asked my chief engineer, Jim Newman, who develops my vehicle systems, to travel to Germany to look her over. Thanks to its research past, *Humboldt* had a lot of what we needed. But it had clearly been outfitted to do more than study the ocean floor. All you had to do was look at the array of antennas. It must have also been a spy ship, used to gather intelligence about Allied navies.

The ship smelled like a toilet—and that isn't a metaphor. There was some sort of problem with the septic system, which could be repaired. The ship was 40 years old, but I told myself it showed promise. *Showed promise?* The last time I used that phrase was when I bought the rundown farmhouse in Hatchville, and it ended up taking 16 years of sweat equity to make it right. *Humboldt* showed promise, too, but I was under no illusions about how much work owning this vessel would entail.

We began a conversation with the broker, who said an American businessman owned the ship. We made a bid of $1.7 million in August 2008. The owner quickly rejected it, but something told me not to give up. Jim asked the broker more about the owner, not expecting to get very far—but, to his surprise, the broker identified the owner as a Wall Street trader named Vincent Viola.

I looked up Vincent Viola on the internet and found he was like me. We'd both been in the military, and we both loved adventure. Vinnie's father was an Italian immigrant who had served in the U.S. Army during World War II. Vinnie graduated from Brooklyn Technical High School and then West Point in 1977. He went to Army Ranger School and served in the famed 101st Airborne Division. He'd founded several successful investment and finance firms and served as chairman of the New York Mercantile Exchange.

We learned that Vinnie had owned *Humboldt* for a couple of years, but now he wanted to sell her. His asking price was a steep $4.4 million—enough, Jim surmised, to recoup his original purchase price and what he'd spent on it since. That was way more than I had to spend—but still, I thought, there might be something to talk about.

I finally managed to find a phone number. I had hardly given my name when the young man on the other end of the line stopped me to introduce himself, saying I had lectured to his class at the Coast Guard Academy. He also let on that Vinnie had planned on converting the spy—er, research—vessel into a family yacht. I asked if I could meet Vinnie, and an appointment was set for noon the next day. We would meet by the clock in the lobby of the Waldorf Astoria in New York City.

That nine-foot-tall clock was a fitting place to meet. Bronze and mahogany, topped with a gold statue of Lady Liberty, it had been a gift from Queen Victoria to the United States, first displayed at the World's Fair in Chicago in 1893. Then John Jacob Astor IV acquired it for his opulent hotel. Nineteen years later, Astor died in the *Titanic* tragedy.

I arrived early at the Waldorf, remembering my grandmother's saying: "If you are not early, you are late, since there is no such thing as 'on time.'" Vinnie, sturdy and fit beneath his military-style crew cut, was prompt as well. Seconds before noon, he walked up and introduced himself. Within moments, we were engaged in conversation at a table in the small dining area.

Vinnie told me he had purchased the ship to bring his family closer. "Our three sons are now grown up," he said, "and my wife, Teresa, and I thought that if we had a boat we could fix up to sail together, it would be a way to ensure we'd be together each year as a family."

Then he asked why *Humboldt* interested me.

I launched into my dreams of exploring the still uncharted regions of the world and beaming our adventures back to shore. I spoke of my passion for ocean exploration and told him about the 16,000 letters I had received from kids after I'd found *Titanic*. With a ship of exploration, I wanted to become a latter-day Jacques Cousteau and inspire children to study science and technology.

When I finished, Vinnie sat there quietly. Then, looking me straight in the eye, he said, "I am embarrassed. Your reason for wanting the ship is more important than mine. I will give you the ship."

For the first time in my life, I was speechless. What do you say when someone does something so generous and completely unexpected? All I could think was, *Thank God he saw the light, and God bless him.*

"Why don't you take over responsibility for the ship right away?" Vinnie said. "My team will draw up a lease agreement so you can get started, and if after five years you still want it, I will transfer title over to you."

Then he added, "But make sure you invite me to go, along with my family, once in a while."

"That I can promise!" I replied. We shook hands on it. As far as I was concerned, a handshake meant everything, and I sensed that Vinnie was brought up the same way.

—

I CREATED A NONPROFIT, the Ocean Exploration Trust (OET), to own the ship and run our new education program. I'd been feeling a bit cramped at Mystic, where my Institute for Exploration sometimes seemed secondary to the larger aquarium operations managed and funded by the board. I planned to stay at Mystic for now and have all my organizations work together. But it felt good to start developing one that was loyal only to me and would give me more independence. Laurie Bradt was one of my first staffers, and Janice Meagher would join us to handle the role that Cathy Offinger had at Woods Hole, taking care of many of the day-to-day tasks for me. Soon I had a new business card that said: "Robert D. Ballard—Explorer—Call Janice."

I also renamed the ship. There was only one choice, of course—*Nautilus*, Captain Nemo's submarine in *Twenty Thousand Leagues Under the Sea*. I chose the prefix E/V for exploration vessel, instead of the normal R/V for research vessel, to drive home a point—that the OET and *Nautilus* were all about fundamental exploration and not specific research projects.

When we found out that the ship's certification as a research vessel was due to expire that December and we needed a million dollars to get her back up to code within a month, I went back to Vinnie and asked for a loan. He said yes, and he wouldn't charge interest. Clearly, he was excited about the prospects for *Nautilus* as well.

I decided to move the ship to Turkey, and I turned to Tufan Turanli to help oversee it. Tufan had worked for George Bass, the marine archaeologist, as director of Texas A&M's Institute of Nautical Archaeology in Bodrum, and he had helped me get the permits for my most recent Black Sea expeditions. He had good relations with the Turkish. I couldn't have done it without Tufan. He remains a great friend to this day.

The ship had plenty of equipment already, but it was still not ready for deep-sea exploration. It didn't have a dynamic positioning system—the ability to hold station, or hover in one place, so that our robotic cameras on their cables stayed where we wanted when we lowered them. It didn't have the multibeam sonar system we needed to map large swaths of the seabed and to find new targets. These upgrades would cost millions, and I knew that was just the start of a drumbeat of massive bills.

I hired Maritime Management, a ship management company based in Dublin, Ireland, to run the ship, which was registered in St. Vincent and the Grenadines in the Caribbean. I also hired 17 Turkish and Ukrainian crew members, and topped it off with a captain who had been an officer in the Soviet submarine force—picture that!

At the same time, I was building a team of young men and women to help me run *Nautilus* and operate its advanced technologies. I had always loved how when Lewis and Clark set out to explore the Louisiana Purchase, they called their team the Corps of Discovery, so I named my team the Corps of Exploration. I decided that women should hold at least 55 percent of the positions on the team, because that's the percentage of women among college students. We need to take greater advantage of what half of our population has to offer, and I wanted these women to serve as role models for the children who would be watching when we rolled out our new education program.

I'm usually pretty calm, even stoic, in the face of adversity, but I didn't get a lot of sleep after I took on that ship. I had a lot of mouths to feed. Many, many nights I tossed and turned and sweated through my options. It was agonizing. For at least the first three years, it was touch and go, financially and emotionally.

I remember thinking, *What other idiots have done this?* And I had to honestly answer that no deep-sea scientists had, because they weren't such idiots. Most of them thought I was crazy. They just looked at me like I was trying to swim across the Bosporus against the tide, or break out of Alcatraz.

Jacques Cousteau had his *Calypso,* but he'd been supported in the 1960s and '70s by revenue from his own diving company and big contracts for TV series like *The Undersea World of Jacques Cousteau*. And he could simply send a cameraman over the side of his ship to film a shark. I mean, how much does it cost to fill your air tank to do something like that? About five dollars.

I wasn't going to be working in the Cousteau layer. We work miles below it, and everything down there is expensive. Need a cable? Oh, there's a quarter million dollars. A winch? Well, that's a million. That multibeam sonar system? Three million, even on a layaway plan.

But I wanted the freedom to explore in whatever area of the deep ocean interested me, and I knew I had to take risks to make it happen. I knew I was exposing myself to failure, and when that happens, there is always self-doubt. But just as in all the tense moments—when *Argo* approached *Titanic*'s bow, or the gear plunged into the sea on the first *Jason* mission—you can never show doubt to the people working with you. Never blink, even though your gut is turning. You need your team to believe in the vision you're trying to make a reality.

—

MY PRAGMATIC GRANDMOTHER would have appreciated my first stopgap solution. I wanted to get *Nautilus* to sea as soon as I could. We needed to generate revenue. But I also needed to find work that we could do as a shakedown cruise, somewhere we could just drop the anchor. That meant working in relatively shallow waters.

We'd decided to base *Nautilus* in western Turkey to use its low-cost shipyards to upgrade *Nautilus*. That area also was almost spitting distance of the site of the naval portion of the Battle of Gallipoli. In 1915, the British had launched a campaign to move ships into the Black Sea through the Dardanelles Strait, intending to link up with Russian forces to fight the Ottoman Empire, Germany's ally in World War I.

Six British battleships were knocked out of commission, including three sunk by sea mines in the Dardanelles, and I was eager to walk this battlefield. The ships had fantastic names, like *Majestic* and *Irresistible,* and they would be easy to reach, lying at a depth of around 180 feet. That gave me plenty of targets. I persuaded National Geographic to fund a TV special on the Gallipoli wrecks. An Australian production company joined the venture, because Australian and New Zealander soldiers who stormed ashore during the battle suffered tremendous losses as well.

I could feel the thrill of heading out for the first time on my own ship, and there was something else special: It was the first time my son Ben, now 15 years old, was joining me on an expedition. We put on our new navy-and-gold *Nautilus* uniforms with a compass patch, and it made our day, father and son together.

Ben had grown up hearing about my searches for ancient ships in the Mediterranean and the Black Sea, and he was enthralled with ancient history. He also had inherited Barbara's family's love for the arts. He had a beautiful a cappella singing voice and a talent for acting. We'd gone on a trip to Greece together, and when I asked where he wanted to go, he knew enough to say he wanted to stand where Leonidas, the Spartan warrior-king, had made his final stand against the Persians in the Battle of Thermopylae. Now he joined other high school and college students as interns on the Gallipoli cruise, and he was having a great time helping to research some of the sunken ships. You can imagine how proud I was when a National Geographic cameraman asked him how it felt to be on the expedition, and he answered, "I've been waiting for this my entire life, you know, going out with Dad, out at sea, and finally I'm here." Like I say, he's a great actor!

As we arrived at the site, I was pushing models of the sunken ships around on a map on the chart table, visualizing what was under us. To make up for not having a good sonar system on board, we towed a separate side-scan sonar vehicle. It would help locate targets, and then *Hercules* and *Argus* would survey this mass graveyard.

The targets showed up as sharp and clear outlines against a field of yellow on the computer screen. We found ships all over the place. *Irresistible* lay in the greenish water, schools of small fish darting over it. We found *Triumph* just a kilometer away from Anzac Cove, named for the Australian and New Zealand Army Corps that fought there. It looked crusted over, with guns poking out, like the other wrecks, and a big gash where a German torpedo had warped the metal.

We also found the Australian submarine *AE2,* which had made it through the Dardanelles and was getting close to Constantinople when it was torpedoed. Many people don't realize how primitive the submarines were in World War I—basically tubes of doom with frequent mechanical problems. Hot, humid, scary, noisy, smelly—I wouldn't ever want to be in something like that, and I've been underneath the sea in a lot of things.

The Germans had given the Ottomans two of their ships to convince them to join their side of the war, and National Geographic wanted us to find the one that had been sunk by a mine. But a storm had kicked up, really bad. I didn't know *Nautilus* that well yet, and I wasn't sure she could handle bigger waves, but I knew I needed to get what Geographic wanted—or give them a pretty good excuse why we couldn't—if I was going to keep the money flowing to fix up the ship.

"Guys, here's what we're going to do," I told the film crew. "You guys get up in the bridge with the cameras. I'm going to put on my foul-weather gear. We're going to put out some safety lines, and we're going to leave our protected area between these islands. We're going to turn the corner and just get the shit kicked out of us, and the waves are going to crash over the bow. And this undaunted explorer is going to work his way up to the bow, just getting thrashed, and we'll tell them that's why we couldn't dive on the wreck. Get ready."

I got all decked out in my gear, looking like the fisherman in the Gorton's fish stick commercials. We turned the corner and I braced, preparing to get beat up. But *Nautilus* was more agile than I expected. She easily cut through the waves, and I was left standing there, looking like an idiot. We still needed the weather to calm down, though, and it eventually did, allowing us to get shots of the German ship, too.

We did a great show, and I was feeling pretty confident about *Nautilus*. She was not just showing great promise—she was making good on that promise, too. Before we flew home, I took CBS reporter Lara Logan out in the Aegean Sea for a profile that *60 Minutes* was doing of me. We were looking for ancient shipwrecks off Turkey, and when she asked me what my favorite discovery was, I answered with the phrase I've used again and again: "It's the one I'm about to make."

Another of my favorite expressions is: You can kick something to death with grasshoppers. Even if each kick doesn't amount to much, they all add up. And it was going to take a lot of grasshopper kicks to get *Nautilus* where I wanted her. The money from National Geographic kept the lights on, and we got *Nautilus* into a Turkish dry dock for the winter. I managed to cobble together enough money from private donors for a dynamic positioning system, but the multibeam sonar would have to wait. NOAA was just launching *Okeanos Explorer,* so I couldn't get funding from them—nor did I want to. I'd be able to get operating funds from NOAA when we returned to basic exploration, but in the meantime, I had to make the numbers work. I went back again and again to some donors, who proved to be patient and generous as I solved the problems with our old spy ship.

One of the key figures who helped me through those early years was Riley P. Bechtel, then chairman and chief executive of the Bechtel Corporation, the nation's largest construction and engineering company. The firm had been founded by his great-grandfather in 1898, and had gained prominence for its role in building the Hoover Dam.

The Bechtel company had been a major supporter of the JASON Project since the early 1990s, and Riley and I had become close friends. We'd vacationed together, and he'd succeeded Don Koll in hosting our annual duck hunt. We were close enough that I wasn't about to ask him to contribute funds for the ship. But one day in the midst of my stress about trying to hit payroll, he turned to me and said, "You know, you've never asked me for help." He insisted that he and his wife, Susie—who was a UC Santa Barbara graduate like me—wanted to make my dreams of exploration with *Nautilus* happen. Riley helped me upgrade the ship, and he continued to step up whenever we had urgent needs. He also vouched for what I was doing, and other people would say, "If Riley's in, I'm in"—including David O'Reilly, chief executive of Chevron, and his wife, Joan, and Wayne Huizenga, founder of Waste Management and AutoNation, and his wife, Marti. In fact, all six of them joined me on *Nautilus* to celebrate at one point.

—

THE OLD GEOLOGIST IN ME was stirring as *Nautilus* sailed south of Cyprus in the summer of 2010. With *Nautilus* upgraded, I could now explore the much deeper waters I'd dreamed of, starting with Eratosthenes Seamount, a strange formation in the middle of the eastern Mediterranean. It brought me back to my days of thinking about plate tectonics.

The island of Cyprus started out as a piece of ocean floor, lifted up above the ocean surface when Earth's African plate crashed into the Eurasian plate millions of years ago. Scientists believe that same smashing of the plates pushed the Mediterranean's seabed higher than the Atlantic's, cutting off the flow of water from the Atlantic and causing the Mediterranean to dry up into basins of salt and exposed dry land. That allowed ancestors of today's elephants to walk across the flat floor between Africa and Europe. Can you imagine that? I fell in love with that whole story of the continents colliding and elephants walking around on the bottom of the Mediterranean Sea.

These same forces produced Eratosthenes Seamount, a flat-topped undersea mountain about 75 miles in diameter. It rises more than 6,500 feet from the sea bottom, and its summit still lies more than 2,200 feet below the surface. It sits beneath the ancient shipping route from Cyprus to Egypt. Eighty percent of all the copper that drove the Bronze Age came from Cyprus, and we all know that Egypt was one of the big kids on the block back then. This was a very rich area to wander into, and I was eager to see what we might stumble upon.

So here I was, sitting in the command center on *Nautilus* as *Argus* hovered over dimly lit pockmarks on the seamount's summit. To our surprise, they turned out to be sinkholes, ranging from 30 to 130 feet across and studded with broken and jagged rock, pale chunks that looked like they'd just been flung around. As we continued on our robotic walkabout, we found dark-stained cracks on the side of the seamount, about 3,000 feet down, with cold fluids seeping through the outcroppings. We saw traces of life. Tube worms swayed, and tiny crabs crawled over limestone formations.

As we descended farther along the slope, we found our biggest surprise. At first glance, it looked like we were driving around a field of salt. What the hell were we looking at? It turned out to be a seemingly endless expanse of thousands and thousands of small white clams, stacked on top of one another. I'd seen fields of clams the size of baseball diamonds at many other hydrothermal vents, but this was mile after mile of clams.

If I hadn't seen it with my own eyes, I would've said it was impossible. Clearly, we'd stumbled onto some kind of new ecosystem that had the capacity to feed these masses of clams. This was different from the hot, nutrient-rich sea vents I was used to. Thermal vents aren't supposed to occur in limestone. The ones I'd found before had fed off chemicals in hot springs bubbling up through Earth's crust, warmed by underlying magma chambers. The seeps were here, on the side of this seamount, but there were no magma chambers underneath. It was too cold for that. So how were the seeps feeding these shockingly large colonies of clams?

Elsewhere on the top of the seamount we encountered what are known as slickensides, deeply grooved fault planes that told us that the limestone cap on top of the seamount was under extreme pressure as the African and Eurasian plates kept grinding against each other. That compression literally squeezes out methane gas, which is converted by resident bacteria into the food that supports the huge population of clams we were seeing.

We also found that the seamount had a lot to reveal about ancient shipping routes. Its summit is basically flat and, given its distance from shore and its elevation above the ocean floor around it, very little detritus falls on it. So if an ancient ship had passed by and threw something overboard, it was still there, thousands of years later. We found hundreds of ancient artifacts, documenting not only the trade route between Cyprus and Egypt, but also trade routes since.

—

MY VISION FOR *NAUTILUS* also included expanding my long-held idea of telepresence for scientists and my new education program for science students to follow our work in real time. Those initiatives got off to a great start on this cruise, too.

As luck would have it, while I was raising money to construct a command center for *Nautilus* at the University of Rhode Island, I had attended a meeting during which National Geographic Society president John Fahey announced he was shuttering the $10 million television studio at their D.C. headquarters and turning it into museum space. I jumped up and said, "I'll take it." My team trucked everything, from monitors on the walls to live transmission gear, to Rhode Island. We called it the Inner Space Center, and it became the relay hub for our two-way live communications between the ship and the outside world.

The Eratosthenes Seamount cruise became the test case for our "Doctors on Call" program, which enabled scientists to view and discuss our video feed remotely to help us understand what we were seeing. Some scientists stood watch at the Inner Space Center, and others followed along through connections at their own universities, on their laptops, or even from their cell phones.

We also geared up a new version of the JASON Project, beaming shows to museums and science centers and streaming them through our new website, *nautiluslive.org*. We had K-12 teachers with us on the cruise, and they and the interns answered questions posed by the kids watching. Over the four and a half months that we explored Eratosthenes Seamount and other areas nearby, some 18,000 questions poured in via the website. It was quite a kickoff, and another way for us to share what we were finding in this rarely seen undersea world. Now we had the template down for future *Nautilus* missions.

At the end of this cruise, as we headed back to Turkey, I felt like I truly had become a more optimistic version of Captain Nemo. I had my own ship, and I was creating a larger world around it.

CHAPTER 15

ALL THAT *NAUTILUS* COULD DO

I t was late June 2012, and I'd gone on a cattle drive with wranglers from the R Lazy S Ranch. It was something I loved to do every time Barbara and I came to the ranch in Jackson Hole, Wyoming, for summer vacation with the kids. My dad had been a real cowboy just to the north, in Montana, and his aunt had married a relative of the famous gunfighter Bat Masterson. So when one of my critics refers to me as an underwater cowboy, I just smile.

My cell phone rang and spooked my horse. I had forgotten it was on. *Must be Barbara calling about dinner,* I thought.

In fact, the call was from Francis J. Ricciardone, Jr., the U.S. ambassador to Turkey. Turkey had severed relations with Syria earlier that year after attacks on Turks and Syrian refugees heading for Turkey. The situation was tense, and the Syrian military had just shot down a Turkish Air Force jet they claimed had entered Syrian airspace. "We need you and your team to find it," Ambassador Ricciardone said.

The Syrians had mistaken the fighter for an Israeli jet, he continued. It had crashed in 5,000 feet of water inside Syria's territorial limits. "The Turks are madder than hell and are mobilizing their army on the border with Syria, and we don't need a war right now. They are convinced that the Syrians have their two pilots. And if they don't, they want you to find them." Ambassador Ricciardone knew of our Black Sea explorations the year before. He had even gone out with us when we were hunting ancient shipwrecks in the Aegean. He knew what *Nautilus* could do.

But I was wary. "What makes you think the Syrians won't sink *Nautilus* if I enter their waters?" I asked.

"We already checked with them," he said to assure me. "They don't want a war either."

I was doing work in the Black Sea under a Turkish permit. If I refused to help the Turks in this emergency, they might not look favorably on my plans to continue running Black Sea expeditions out of Turkish ports. "What about insurance?" I asked. The Turkish government was going to pay for the expedition and guarantee coverage, Ricciardone said.

I told him I needed to call Vinnie Viola, who still owned the ship under our lease agreement.

Vinnie's response was immediate: "Can I go with you?"

"That's up to you," I said.

"When do I pick you up in my jet, and where do we go?"

Just like that, my cattle drive was over.

I called Katy Croff Bell, who was then my second-in-command at the Ocean Exploration Trust and the lead scientist on *Nautilus*. Katy has a bachelor's degree in ocean engineering from MIT and a Ph.D. in geological oceanography from our program at the University of Rhode Island. She was about to fly from Boston to Turkey to start up our exploration season, including returns to the Black Sea and Eratosthenes Seamount planned for the summer. Now we had a bit of a detour before we could get on with that cruise.

I told her what was going on and said she should get *Nautilus* under way as soon as possible. Head toward Syria, I said, but stay in Turkish waters. I asked Tufan Turanli to organize a tour of ancient ruins there for the students, teachers, and scientists already gathering in Turkey to join our expedition. We needed to occupy them while we took care of the situation, and then we would return to pick them up.

I met Vinnie at the Teterboro, New Jersey, airport—the main tarmac for the private-jet set in the Greater New York area. I was making good on my promise to take Vinnie on an adventure—but not the kind either of us had envisioned.

We arrived in Turkey on July 3. A convoy carrying a Turkish diplomat and military officers took us to the frigate T.C.G. *Gokceada* to set off for our rendezvous with *Nautilus*. The Turks briefed us as we sailed, and I made clear that if they wanted me to find the jet and recover the bodies, they needed to tell me the truth about what had happened.

They told us that on June 22, Turkish radar operators had noticed that one of its American-made F-4 Phantom jets was experiencing "unbalanced movement" 13 nautical miles west of Syria. That was when it was hit, the Turks assumed. The jet was headed due east and was only seconds from entering Syria's territorial waters, which stretched 12 miles out from the coast. Flying in that close was playing with fire. The jet disappeared from radar two minutes later. If it had crashed, it was certainly within Syrian waters.

An hour and 40 minutes after that, a Turkish helicopter spotted an oil slick, and a Syrian warship noticed debris floating nearby, including the pilots' flight helmets, boots, and survival kit. This is what got the Turks worked up. How could Syria have taken possession of the pilots' gear but not the pilots? Could they still be alive?

When we reached *Nautilus,* the Turkish officials took Vinnie and me over to her on a launch. With escort ships and helicopters circling around, the two ships entered Syrian waters to begin the hunt. Both Turkish and Syrian warships stood guard, eyeing each other warily but knowing they had to cooperate.

I had a terrific team of young men and women in my Corps of Exploration. The summer before, as I directed the operation from our command center in Rhode Island, they had found a ship from the third century B.C. off the Turkish coast, with visible leg bones of one of its sailors. The wreck was 330 feet underwater, so shallow that the water was saturated enough with calcium carbonate that it did not leach it from the bones. My team had knowingly signed up for scientific expeditions like that, but not for service in active war zones.

Just 10 days before the jet was shot down, the United Nations had formally proclaimed the fighting among factions in Syria to be a full-blown civil war. We were close enough to shore that we could hear artillery rounds and see the smoke. So yeah, that was intense.

The Turks trusted me, and with my permission, gave Vinnie access to everything we were doing. If I hadn't known about his Army Ranger background, I might have been more concerned by one of his habits: He crosses himself frequently. But Vinnie is a seasoned soldier as well as a deeply religious Catholic, and his presence helped keep our young crew calm.

I had an abundance of data to work with—more, in fact, than on any search effort I had ever conducted. A Turkish ship had made a sonar map of the ocean floor, identifying eight possible targets. My team added the locations of the floating debris to determine drift patterns, which we could use to point back toward the impact site.

One of the targets had the rough dimensions of the jet, and the Turkish officers wanted to check it first. I was skeptical. I knew from the way the debris had drifted that it couldn't be the right spot. We launched *Hercules* and *Argus* and determined that it was just a sunken shipping container.

I selected the next target, and we towed the vehicles over to it. It turned out to be one of the jet's engines, lying by itself. Soon, more pieces of the wreckage came into view.

For hours, I sat fixed at the console, joined by Yaşar Kadioğlu, a Turkish Air Force colonel who wore a jumpsuit covered with military patches and a silk scarf. At first he reminded me of Iceman, the pilot in *Top Gun* who was completely full of himself. But I would come to know another side of Kadioğlu and to see how dearly he loved his pilots.

We stayed at the console for hours and hours. Our meals were brought to us, and we pushed forward on pure adrenaline. The colonel kept a running tally of the objects we were seeing: "Landing gear . . . electrical generator . . . engine fan . . . second engine . . . rear section of the cockpit . . . part of the fuselage."

He asked us to zoom in on that piece of the fuselage. The Syrians claimed to have shot down the jet with antiaircraft fire. That would have left numerous holes in the shiny metal, but there were none. The colonel didn't say anything, but it was clear to us both that the jet had been blasted into so many fragments that it had to have been hit by a missile.

Just south of the second engine, we came upon the control panels from the cockpit. We hovered over that area and brought up the panels. The dials were frozen in time—the altitude, compass bearing, and speed reading out exactly as they had when the jet was vaporized.

If there were bodies to be found, we must be getting close.

"Do you see that line of crabs on the bottom?" I asked Brennan Phillips, my *Hercules* pilot. "Head in that direction."

My mind drifted back to 1969, when I had just left the Navy and was working at Woods Hole, making my first dive ever in a deep submersible. We had dropped a 55-gallon drum full of dead fish to the bottom to see what creatures it would attract. When we returned to check on it later, long lines of crabs were heading toward it, and the drum itself was lost from view, overtaken by a giant pile of crabs.

It is never a pleasant sight to find someone who has died in an accident, but it's worse when other creatures have made that discovery before you. The crabs proved to be reliable direction finders, leading us to the bodies of Capt. Gokhan Ertan and his copilot, Lt. Hasan Huseyin Aksoy.

We began to assemble the elevator we use to recover scientific samples and archaeological artifacts. We had used it many times before, but this was the first time it would be used to recover bodies. We moved *Nautilus* into position to drop the elevator, letting it drift to the bottom. Cagatay Erciyes, deputy director general of maritime and aviation affairs for Turkey's Ministry of Foreign Affairs, joined me at the rail to watch.

"You know, Dr. Ballard, it was very difficult for us to make the decision to ask America to help us with this effort," he said. "We are a very proud nation and should have been able to do this by ourselves." He paused for a moment, watching my team in action. "I would like to ask if you would be interested in selling us your ship and technology."

I thanked him for the offer. "It's not just the technology, but my team, that makes all this possible," I told him. I pointed to Chris Roman, a Ph.D. from the joint Woods Hole–MIT program, who was operating the crane, and to Katy Croff Bell, our lead scientist, who was handling the launch lines attached to the elevator. "They are both priceless," I said, implying, No, *Nautilus* is not for sale.

It took 90 minutes for *Argus* and *Hercules* to reach the bottom. Brennan Phillips, who would earn a Ph.D. in oceanography from the University of Rhode Island, was driving *Hercules,* controlling its manipulator arms to move the elevator next to Lt. Aksoy's body. Then he had the supremely delicate job of maneuvering the body into the hammock. Guided by the video feed, Brennan carefully made *Hercules* grasp a strap on Aksoy's flight suit, lift the body just enough to get it into the metal hammock, then pull the pin that closed the mesh lid over the body.

Before we could release the weights and let the elevator float to the surface, we had to be sure the Syrian and Turkish warships got out of the way, so the elevator didn't come up under one of their hulls. Colonel Kadioğlu got on the radio and asked them to move at least a mile away. Both complied immediately, a cooperative pattern that held throughout the mission—strong evidence of how much both sides wanted to avoid conflict.

Brennan then sent a signal that released the weights. We could see the copilot's body, secure in its hammock, as the elevator floated straight up and out of the view of the ROV cameras. Its yellow floats reached the surface, and two Turkish Navy divers secured the body. With a thumbs-up from them, we lifted the hammock out of the water.

We then repeated the procedure to bring up the body of Captain Ertan, the pilot.

It was a solemn ceremony. The most moving moment of all was when Colonel Kadioğlu approached the bodies we had brought back on board. He reached into the top pockets of each pilot's flight suit and carefully removed pictures of their wives and children, placing them in plastic bags to return to their families.

The recovery complete, the Turkish warships formed a tight circle around *Nautilus*. Their crews stood ceremoniously on their decks, lined up from bow to stern, wearing dress whites and standing silently at attention as the ships' horns saluted their fallen comrades.

A Turkish naval physician supervised the transfer of the bodies to *Gokceada*. Vinnie and I were asked to escort them onto the Turkish vessel, passing by an honor guard of Turkish sailors. After a ceremony on the fantail, the first helicopter carried the bodies back to the mainland. Vinnie and I left on a second helicopter, which vibrated so much that Vinnie later said it scared the hell out of him—it sounded like the rotor blades were going to come flying off. It hadn't bothered me—I tend to be fatalistic in these situations. We made it to shore, where I briefed the area commander of the Turkish fleet. Mission accomplished.

The Turks made good on their pledge to pay for our mission, inviting us to fill *Nautilus*'s huge fuel tanks at one of their docks, arranging our flight home, and eventually wiring us a welcome $750,000. *Nautilus* and the Corps of Exploration became heroes in the Turkish news media, and we were treated very well on our flight back.

Shortly after, I found myself sitting in a secure room at the submarine base in New London, Connecticut, a short drive from home. Adm. Jonathan Greenert, with whom I had sailed on *NR-1*, was now the chief of naval operations, and he wanted a briefing on the mission. I also went to the Pentagon to meet with the vice chairman of the Joint Chiefs of Staff, who wanted to know what kind of missile brought down the Turkish jet. I knew it was most likely a Russian missile, and I shared my photos of the wreckage. And that was the end of that. I've done many briefings like this over the years, and every time it's a one-way street. I hand them everything I have, and they just say, *Thank you very much; we'll call if we need you.*

One final reminder of the military nature of our mission came several weeks later. Three members of my team asked to take leave from *Nautilus,* haunted by what they had seen. I suggested I would connect them with a psychologist who worked with military personnel returning from war zones, but they all passed on the offer and returned to sea later that season.

—

BEN CAME WITH ME AGAIN that August for our second exploration of the area around Eratosthenes Seamount. We wanted to use *Hercules'* high-definition still and video cameras to take a closer look at the sinkholes, fluid seeps, and vent fields, not to mention all those clams.

The lack of sedimentation meant that history had just been dropped on that tabletop. We had found 79 amphoras on the seamount's flat top in 2010, and we found 149 more in 2012, mostly dating from the sixth century B.C. to the seventh century A.D. We also found a shipwreck that seemed to date to the fourth century B.C.

After studying the pictures of the artifacts, Ben said we had found an even older amphora. "I think it's Canaanite," he said, referring to the Bronze Age civilization that had existed along the eastern shore of the Mediterranean.

"But Ben," I said, "that's 2000 B.C., for Christ's sake." It seemed farfetched, way too old.

"Dad, I'm sorry, but it looks to be Canaanite," he insisted.

I didn't believe it. First off, very few intact artifacts from the Canaanites exist anywhere. Second, Ben wasn't an archaeologist or an expert on amphoras. But he had been staring at them with me for a couple years now, and I didn't want to argue with him. I told him to call Larry Stager, who'd searched for Phoenician ship-wrecks with me in 1999.

Larry's response when he saw our images via telepresence? "Your son's right."

That really impressed me. It also showed that even that far back into time, mariners were sailing far from shore.

We had 12 K-12 teachers on that cruise with us that summer. They hosted more than 300 shows, which we beamed live to museums and science centers. We streamed our explorations live through three different video feeds, and the teachers answered more than 11,000 questions coming in through our website. They worked brutal hours, compared to a normal school schedule, but some were already asking to come out again.

It felt as if we'd hit a home run with this expedition. Ben used our Eratosthenes research for his first appearance as a senior author of a paper in a scientific journal, tracing the shifting powers in that area, going back to the time of Troy and the Greeks. Ben and his co-authors used a color-coded chart that showed the trails of amphoras discarded by ships crossing the Eastern Mediterra-nean, documenting ancient trade routes crossing over Eratosthe-nes from Greece, Egypt, Cyprus, Rome, and the Near East, along with the time periods in which those crossings were made. From that, they could determine changes in the power structure of the region based on shifting patterns in maritime commerce. We didn't even have to pick up the amphoras. Ceramics experts could just look at our photos and find the telltale signatures.

The whole exercise helped me realize another way in which I differed from traditional marine archaeologists. When they found an ancient shipwreck, they focused on excavating and document-ing every inch of the ship. My reaction to that approach was just, "Oh, my God, really? You're going to spend your whole life on one ship?" I was more interested in the big picture. Where were they going? What were they carrying? The commerce mattered most, because that's what spread ideas and knitted worlds together.

—

THIS AREA AROUND THE SEAMOUNT was a gold mine. I could have camped out on this puppy for the rest of my life. But I had to walk away from it, and that hurt.

I rarely give up on anything I care about, and I'm persistent enough that I can usually wear down any resistance I run into. But two things were happening that forced me to alter course. First, congressional priorities were changing. Lawmakers wanted NOAA to quit funding international expeditions and focus on U.S. waters. Second, tensions with marine archaeologists were heating up again. My goal was to develop technology to excavate ancient deepwater shipwrecks to archaeological standards. But the archaeologists I was working with could not raise the funds from their own community of sponsors to make that happen, and some of them wanted to be in charge even though my team was doing the work. I wasn't happy about either prospect. I was enthralled by human history. I loved the Greeks, with their philosophy of sound mind and body. We'd just started working our way into Greek culture. And I still wanted to explore the Indian Ocean. But the rise in piracy had made it impossible for us to get there safely, and now it looked as if I wouldn't get the chance.

So if I had to stay closer to home, it looked like I should shelve my archaeological oceanography program and save myself the stress.

I understood the political and economic imperatives under which NOAA was operating. I was now serving on NOAA's Science Advisory Board, and I had previously served on the presidential commissions. I had heard the recommendations that funds be used primarily to explore America's exclusive economic zone

(EEZ), the band of ocean 200 miles off our coasts, where the United States held the rights to the mineral and fishing resources, as demarcated in the 1982 United Nations Convention on the Law of the Sea. *Okeanos Explorer,* the ship I'd help NOAA obtain, was mapping the EEZ in the Atlantic. Up until now, that had been sufficient.

But after the Republicans took control of the House in early 2011, some of them had pushed NOAA to start mapping the EEZ in the Gulf of Mexico and the Pacific as well. I depended heavily on NOAA's grants to keep *Nautilus* at sea. There's a great saying that you can't herd cats, but you can move their milk. In my business, that milk is money, and the government was herding this cat in a new direction. If a super-billionaire had come up to me and said, "Hey, the hell with them, I'll sponsor you," I would have been back over the Eratosthenes Seamount and into the Indian Ocean in a heartbeat. But no billionaire stepped in, and I had to go where my funder wanted me to go.

The situation with academic marine archaeologists was harder to swallow, after all my efforts to bring archaeology and oceanography together into a joint field. I got along with some of the best marine archaeologists, like George Bass, as well as with great land archaeologists like Larry Stager and Fred Hiebert, but it seemed like others just wanted us to use our money to find ancient shipwrecks and then step aside with not so much as a thank you. Some treated the scientists and engineers on my team, people with Ph.D.'s from MIT, like technicians, not colleagues. A couple of archaeologists actually published a paper calling for a committee of archaeologists to approve or disapprove all deep-sea expeditions like mine. It was also turning out to be difficult to attract enough students into our Ph.D. program at Rhode Island. Though we'd graduated some terrific students, we couldn't find enough who knew ancient languages and history and also could cross the divide into high-level science. Archaeologists and physical scientists, it turns out, are cut from very different cloth.

It was time to move on. So in my glass-half-full way, I decided to embrace my new assignment and push it to the max.

—

OUR JOB, much like the first Lewis and Clark expedition, was to map unknown territory, this time all those swaths of sea up to 200 miles out. In fact, the United States owns more land underwater than any other nation on Earth. I call it the "New America," and it equals the mass of our terrestrial landholdings. Yet we have better maps of Mars and the far side of the moon than we have for 50 percent of our own country.

Our new plan was to spend the first two and a half years in the Gulf of Mexico and the Caribbean, including the EEZ around Puerto Rico, working our way toward the Pacific Ocean, where we would set up camp for the long haul, because that's where most of the unmapped and unexplored portions of the U.S. EEZ are located. We mapped the Caribbean and Gulf seabed in 2013 and 2014. That gave us several new exploration opportunities. We examined hydrothermal vents in the Mid-Cayman Rise south of Cuba, explored the geology in the Straits of Florida for evidence of tsunami-generating landslide events, and inspected the wrecks of three early 19th-century ships off Galveston.

We kept expanding our education programs, too, streaming these explorations live on our website, and we did some freelance work to bolster finances. We took other scientists out on *Nautilus* to help them study the impact of oil and gas seeps on deep-sea coral communities and the harmful effects of the dispersants that had been used to help clean up the 2010 BP oil spill. In the spring of 2014, I decided to leave the Mystic operation altogether, bringing Laurie Bradt, Dennis Armstrong, and Janice Meagher to help me run the Ocean Exploration Trust. It was now headquartered in an old granite mansion across the street from the Coast Guard Academy in New London, just down the road from Mystic.

I also got interested in a bit of naval history called Operation Drumbeat, Adolf Hitler's program to send U-boats to our East and Gulf Coasts in an effort to move World War II into U.S. territory. Shortly after the Japanese attack on Pearl Harbor and the German declaration of war against the United States, Germany directed five of its most advanced U-boats to attack American shipping along our coasts. In less than a month, in early 1942, U-boats sank more than 150,000 tons of cargo without suffering a single loss. Over three and a half years, they sank 397 ships and killed some 5,000 people, but only one German U-boat was known to have been sunk in the Gulf of Mexico during the operation.

It was all interesting enough that in July 2014, we filmed a National Geographic special, to be broadcast on PBS's NOVA series. Titled *Nazi Attack on America,* it explored Operation Drumbeat and the wrecks it left behind. We started with two oil tankers that torpedoes had sunk, but we focused especially on the remnants of a Nazi submarine, *U-166,* trying to solve a mystery that still puzzled the U.S. military.

On July 30, 1942, *U-166* crept up on the American passenger ship *Robert E. Lee* as it steamed toward New Orleans. The Germans fired a torpedo, a direct hit, and *Lee* sank in less than 10 minutes.

Just after the steamship was hit, the crew of a nearby escort patrol boat, U.S.S. *PC-566,* saw a periscope in the water. Capt. Herbert Claudius reported that he turned his boat toward the periscope, judging where he thought the German sub would be. He snuck up on the U-boat and dropped five depth charges on his first pass over, then turned around and dropped five more on a second pass.

Claudius saw an oil slick—evidence to him that he had hit the sub—and he reported that the U-boat had gone down. But his superiors did not believe his report. He'd never had antisubmarine training, and he had set his depth charges to go off at a much deeper depth, though the sub was near the surface. There was no way he could have sunk it, they concluded.

Two days later, a Coast Guard plane reported that it had dropped a depth charge and sunk a diving U-boat 130 miles from where *Lee* had gone down. With that information, the military considered the case closed, presuming that the U-boat had slipped away after Claudius's attack and that the plane had hit it later. Claudius was reprimanded for having broken radio silence earlier in the patrol, removed from command, and sent to anti-submarine warfare school.

That's where everything stood until 2001, when an oil field contractor happened upon the wreckage of a U-boat just one mile from where the *Robert E. Lee* had sunk. That alone seemed to confirm that Claudius had been right, but the Navy had never put in the time needed to correct the record. Had the wrong unit been credited with sinking the sub? Could the technology we had aboard *Nautilus* help exonerate Herbert Claudius?

Lara Logan of *60 Minutes* was with us again, filming a piece on my push to explore U.S. waters. Also aboard with us was Richie Kohler, the shipwreck diver famous for his dogged search for another U-boat that was described in the best-selling book *Shadow Divers.* We had solid information about where both *Robert E. Lee* and the U-boat sank, so we headed to those coordinates and sent our trusty robots down to take a look. Lara, Richie, and I watched the computer screen together as *Hercules* crept along the bottom.

Robert E. Lee seemed to be in decent shape, with the windows battened and a bell attached. A gun, symbol of the dangerous times in which the ship went down, was still attached to the stern but was covered in purple and orange sea anemones. Orange-topped anemones stuck out all over the bow like wavy mushrooms.

Then the U-boat came into view. It was dull green, covered in silt and overwhelmed by anemones, just like the *Lee*. Richie instantly started pointing out features that showed that the sub had faced a perilous situation and submerged quickly. Its antenna was bent, and its periscope had never fully retracted. We could see that it had broken into two pieces, with one section out of view.

We sent *Hercules* along the empty seafloor to find the missing section. A gun came into view, which amped up excitement in the monitor room. Eventually the missing part of the ship, its bow section, came into view. We had found what we were looking for.

Hercules swept over the bow and the rest of the sub, taking more than 2,000 pictures, which we would use to create a photomosaic of the entire sub to see what more we could learn.

I already felt we had enough evidence to right the wrong done to Captain Claudius. The U-boat sank right next to the steamship, just as Claudius had reported.

I sent an email to Jonathan Greenert, who was still the Navy's top officer—and a friend.

"I am presently aboard the E/V *Nautilus* in the Gulf of Mexico," I wrote, and described what we had just found. "Not only was Captain Claudius never credited with sinking *U-166*, he was reprimanded, relieved of his command . . . As a result, he went to his grave never having been honored for sinking *U-166*."

I ended my note with a request. "Do you think it might be possible to correct this wrong?"

Admiral Greenert soon responded, telling me he'd look into it. At his request, I wired our findings to Navy officials, who assigned several maritime experts to dig through the historical record.

Once we got back to the University of Rhode Island, our mapping specialist, Claire Smart, put together a photomosaic of the sub. She spent hours and hours on the computer, looking at the indistinguishable gray piles on the screen and piecing them together by eye.

When she finished, I invited Richie Kohler to the lab to take a look. We spread the printed photomosaic out on the table, comparing it with an image of the sub before it sank. We focused on the main break in the hull. It was right at the forward torpedo room, and we could see that something had destroyed the top of the room. What if one of the 10 depth charges had landed on the deck and stayed there as the sub crash-dove to escape its pursuer, then exploded above the torpedo room when it reached the depth Claudius had set? That would have provided more than enough explosive power to blow off the bow of the sub.

Kohler also found declassified World War II cables and other records that supported Claudius's story. It turns out that two U-boats were in the Gulf at the time, and the captain of the second one had reported that he had evaded an attack by a U.S. plane. That detail explained the claim of the Coast Guard pilot and left only the sub we had just inspected, *U-166,* unaccounted for.

The Navy confirmed our re-creation of the scene and decided to honor Herbert Claudius posthumously. I attended a beautiful ceremony at which Admiral Greenert, the chief of naval operations, and Ray Mabus, secretary of the Navy, awarded Gordon Claudius, the captain's only surviving child, the Legion of Merit on his father's behalf.

I stood next to Gordon during Mabus's speech, listening as the secretary said, "It's never too late to set the record straight. It's never too late to do the right thing."

Amen to that. There wasn't a dry eye in the room. It was a good day, knowing our team had helped restore someone's honor.

CHAPTER 16

DISCOVERING MYSELF

Writing this memoir has given me the opportunity to reflect on my life and character, the stories I can tell and the lessons I have learned through my 70-plus years on this planet. What is it that made this kid, born in Kansas and raised in Southern California—this kid who had trouble reading and couldn't channel his energy until he started dreaming about becoming a modern-day Captain Nemo, this guy who stumbled and fell more times than I can remember and picked himself up and kept on going—what was it that got me to where I am today?

I ask myself these questions as I look around every June, taking part in National Geographic's annual Explorers Festival. Wherever I have been, whatever projects I might be wrestling with, I always find my way back to Washington, D.C., for the presentations at Society headquarters and evenings of storytelling with some of my friends across the street at the Jefferson Hotel. We're only five blocks from the White House, and so the elegant wood-paneled bar is filled with political types. It's become a regular haunt for us as we reunite year after year.

Dereck and Beverly Joubert stop by. They film Africa's big cats—lions, cheetahs, and leopards. Dereck looks the part, with his long mane of white hair and habit of wearing neck scarves. Beverly had a terrible accident in Botswana in 2017. A Cape buffalo charged out of the darkness, knocking Dereck aside and impaling Beverly just beneath the armpit. Dereck managed to fight off their attacker, suffering broken ribs and a cracked pelvis. Beverly lost five pints of blood, but one of the first things she asked her surgeons was when she could get back to work. Talk about courage—and passion.

Lee Berger is always at our gatherings. Lee electrified everyone in 2008 and 2013 by finding new species of hominins in South Africa—not one but two different species on the human evolutionary tree. He advertised on Facebook to find archaeologists, mostly women, who were intrepid and tiny enough to wiggle through rocky cave openings seven inches wide to get to a deep cavern full of fossil remains. As he points out, the places left to explore are getting more remote and dangerous.

Enric Sala always joins us, along with his partner, Kristin Rechberger. I love the twinkle in his eye. Enric, a marine ecologist who runs National Geographic's Pristine Seas program, studies the parts of the ocean likely to be least affected by civilization. The Pristine Seas Project has been instrumental in convincing governments to establish more than five million square kilometers of protected marine reserves. Enric reminds us that these days explorers cannot just be observers. We spend our lives investigating amazing natural places, and it's our responsibility to figure out how to protect them in the face of threats like climate change and the growing world population.

Then there is Kenny Broad, an environmental anthropologist and cave diver. Kenny's crazy. He likes to joke that he goes exploring in muddy holes. But what he really does is squeeze his way through dark underwater caves, researching some of the least accessible environments on Earth. I wouldn't do that. I can handle claustrophobia, but being buried alive, no thanks.

One of my greatest joys is having conversations with friends working in other disciplines that can make me think more deeply about what I'm doing and send me off in new directions. Kenny Broad and I are working together on a project we call Walking with the Ancients. It's a good example of how exploration is changing, and it's about as collaborative as you can get. The program was inspired by the work of Spencer Wells, the geneticist who founded National Geographic's Genographic Project. Spencer used his research into human DNA to create what he called a map of our wanderings—the migration routes of early humans as they walked out of Africa and through other parts of the world.

Spencer's research showed how humans had crossed from Siberia into Alaska during the last ice age, some 20,000 years ago, when rivers became glaciers, locking up Earth's water, dropping sea levels, and creating new shorelines. Today's sea levels are 400 feet higher than they were during the last ice age. Along the Southern California coastline where I grew up, there could be an underwater world that our human ancestors knew but that we have yet to discover. Researchers have presumed that during the southward migrations that populated the Americas, our ancestors either canoed around the glaciers or waited several thousand years in the far north for an ice-free inland corridor. But why migrate in canoes if they could walk along the beach?

I remembered my first scuba dives as a teenager, when I found a cave full of lobsters 33 or so feet below the ocean surface. This cave, and deeper ones I proposed to explore, would have been above water 20,000 years ago. Early humans might have sheltered here. Recreational divers have explored and scavenged the shallower caves, but any caves deeper down likely remain untouched. The first step was to prove that the caves existed.

Nautilus was already in the Pacific, mapping the seabed in the exclusive economic zone there. So we started searching for caves along the way, too, using both our robots and Kenny's divers. Each day we'd wake up at five, and the divers would put on their game faces and go. Ocean surges can rush into these caves, filling the divers' ears with pressure and compressing their chests. It's dark, too, so each diver carried flashlights and extra-thick safety lines that wouldn't break if a surge smashed them against rocks.

The caves were right where we expected them to be, in several tiers down to depths of nearly 400 feet. Kenny and his divers explored the ones about 230 feet down, right at the edge of where they could safely work, grabbing two quick sediment samples. And what do you know? Our genetics experts determined that human DNA was in one of them.

It's looking like my hunch was right—some early humans may have camped in those caves on their way down the coast. We plan to send *Sunfish,* an autonomous underwater system, to explore caves in the three deepest ancient shorelines, from 328 feet to almost 400 feet deep. Who knows what else we'll discover? Will there be shards of pottery? Other signs of human use? We don't know, but we're ready to find out.

The Walking with the Ancients project is just one example of the synergies that come out of our chats during the Explorers Festivals. The gatherings remind me of stories my father shared from his Montana boyhood. Fur trappers would emerge from the wilderness by the hundreds each year to camp on grassy plains in Wyoming. They would bring their furs and hides to sell and replenish their supplies for the long winters ahead. These were mountain men, so there was a lot of drinking, knife and tomahawk throwing, and storytelling about living in the wild.

At our meetings, there's not much knife throwing, but there is a lot of storytelling and drinking late into the night. Exploring is a lonely pursuit, and it's exciting to be surrounded by others who have stared into the unknown. Many of us did not feel completely comfortable in the families or societies we grew up in, and that makes it easier to leave civilization and head into the wilderness.

It's also a thrill to get to know the younger explorers at the festival, as they talk about using sensors, satellite links, and machine learning algorithms to monitor the health of penguins in Antarctica or praying mantises in Brazil. To many of them, I'm "the *Titanic* guy." They want to make their mark and get to the point where my friends and I are, with the freedom to go out and explore whatever interests us. It makes me feel like a tribal elder when I share what I've learned with them.

—

I WONDER WHAT those younger explorers would think if they knew that I've always felt insecure, deep down, about my brains and ability. Even with all the success I've had, I've never been able to shake the fear that I was like a boxer trying to fight opponents in a higher weight class.

A lot of it has to do with my boyhood feeling that I didn't measure up to my brother, Richard. My difficulties in reading were a challenge, but I developed work-arounds, like spending more time on my homework. Once my career got going, I was always in motion, barreling from one project to another. By then, I believed that with hyperfocus and energy, I could overcome any challenge. And although being a more visual thinker helped me in my science and shipwreck searches, I never understood why I seemed to think differently from others around me.

Then, one day in March 2015, I was driving home from my office, and I heard a segment on the radio about a book called *The Dyslexic Advantage*. What I heard felt familiar. Could I have had dyslexia without even realizing it? I ordered the book that night, and when I began reading it, I couldn't put it down. Tears were streaming down my face. Here I was, 72 years old, and this book, finally, was explaining me to me. Even now I think of it as my first autobiography. That's how much it mirrored who I am.

In their book, authors Brock L. Eide and Fernette F. Eide, both M.D.s, make the point that although the brains of those with dyslexia are wired differently, their thinking isn't defective, just different—and a gift. A gift? I'd never thought of it that way. The Eides also noted that John Lennon, Charles Schwab, and Richard Branson were all dyslexic. *Who wouldn't want,* I thought, *to join a club like that?*

The Eides identified four patterns of reasoning in which those with dyslexia excel: material, interconnected, narrative, and dynamic—a foursome conveniently shortened to MIND. I find pieces of all of them in myself.

"Material reasoning" refers to the ability to see and work in three dimensions, recognizing shapes and sizes and how they interact with each other. People with dyslexia strong in this type of reasoning gravitate to fields like architecture and engineering. I'm not an engineer, but I work closely with them and feel comfortable thinking about objects in space. I'm thinking about the dicey moments, like the first time we took *Argo* over *Titanic,* when I was visualizing the obstacles we needed to avoid. I'd take in all the displays from our sonar sensors and cameras—depth, range, shape, color, everything else—and create a mental 3-D picture of the undersea terrain. It was pitch-black beyond our camera lights, but I knew what was out there. It's like the lights came on in my head.

"Interconnected reasoning" is just a fancy term for interdisciplinary thinking, or the ability to mix and match different ideas or approaches in novel ways. As you've seen, I've never paid attention to strict intellectual boundaries, like the differences between oceanography and archaeology or Army and Navy tactics. My thoughts floated from one to the other. I've always seen the benefit of grabbing knowledge wherever I could and using it to create new approaches. Knowledge doesn't fall into neat, discrete categories, at least not in my world. I also love to use analogies, like calling myself a Swiss Army knife to describe how I became a jack-of-all-trades.

My love of analogies also helps me with "narrative reasoning," which refers to the inclination to tell stories not only to transmit information but also to comprehend it. Rather than thinking in abstract terms, people with dyslexia collect concrete examples and combine them in creative ways. When I give lectures, I don't write anything beforehand. I never use a teleprompter. I just call up a slide with an image to cue the next story. Ask me a question, and I'm more likely to relay a scene than state a fact. When Vinnie Viola asked me why I wanted a ship, I told him about the 16,000 letters kids wrote me after I found *Titanic*.

"Dynamic reasoning" is the ability to see patterns and use those insights to reconstruct the past or predict the future. That is my strongest card. My debris field theory grew out of seeing Jack Thayer's drawings of *Titanic* splitting in half and imagining how the debris scattered from there. It's about action; it's about doing it. My big breakthrough as a scientist was to turn *Alvin* into my field jeep and go deep to observe the seabed myself rather than relying just on sonar echoes. By the time others started using that technique, I had moved on to robotic cameras. After I found *Titanic* and cute robots were becoming all the rage, I was pushing to get funding for telepresence, so students and scientists who never left their armchairs could participate in our discoveries. It took 10 to 15 years for each change to gain full acceptance. That's why I like to say that crystal balls were probably invented by someone with dyslexia.

The Dyslexic Advantage caused all these thoughts to start tumbling out. I looked back over my whole life and saw things in a new way. I grew up in the late 1940s and early 1950s, and no one was talking about dyslexia then. Researchers today estimate that anywhere from 5 to 17 percent of the world's population may have dyslexia. That's several hundred million people. But back then, kids who had trouble keeping up in school were just considered slow learners.

Dyslexia is widely recognized today, and teachers can test for it if red flags show up. They have gotten better at working with kids who have dyslexia, but I wish everyone recognized that dyslexia has advantages, and the best education is one that brings those talents out in children.

Research is showing that there are genetic connections to dyslexia. So where did mine come from? My mother was an excellent speller. She was my dictionary growing up. My dad, on the other hand, always had trouble spelling. Once he came to visit me at work at Woods Hole. That day, for some reason, everyone on my team was wearing a red shirt. My dad's shirt was brown. He took out a magic marker and wrote "Read Shirt" on it. Even now, if I was asked to spell "restaurant," I couldn't do it.

I look at the nature versus nurture question this way: Your genes are the nouns and the verbs, and your nurturing is the adjectives and the adverbs. Together they make a sentence, with your genes, like all nouns and verbs, having the strongest impact.

Dyslexia and ADHD, attention deficit/hyperactivity disorder, often come together, and I've thought about that as I reflected on my life as well. To this day, my biggest challenge is to manage my subconscious, which never stops thinking, 24/7. It wakes me up at night with something, and I'll often tell Barbara that I got up in the middle of the night to deal with a monster.

I've learned to handle it during the day through the time management and discipline my mother taught me. And when my brain gets too revved up, I stop and do something physical, like chores around the house or tending the garden. I also find that doing jigsaw puzzles, even difficult ones with no helpful picture on the box, gives my mind a chance to calm down and think clearly again.

My first two sons, Todd and Dougie, were hyperactive growing up. I put a sign on Dougie's bathroom mirror: "My body is like a race car, and when I learn how to drive it, I'm going to win lots of races." As a matter of fact, Dougie learned to channel his energy so well that he now runs a successful business. My son Ben is only mildly dyslexic. He reads well and recently earned a master's degree from the Fletcher School of Law and Diplomacy at Tufts University. My daughter, Emily Rose, has dyslexia, and she initially had a hard time fighting through all the labels that suggested she had a learning disability. She attended a small private school with lots of superachievers with their eyes set on the Ivy League, and she was bullied. But Emily has the strongest moral rudder I have ever seen. She resisted any kind of help with her schoolwork, viewing that as cheating, but Barbara finally found a tutor who had a child with dyslexia. After Emily embraced her dyslexia and began to discuss it openly, she began to shine. She loves a quote from Walt Disney—"If you can dream it, you can do it"—and she gave me a little sign with it that I keep at my desk.

My mother died of pneumonia at the age of 98 in August 2015, just five months after I discovered *The Dyslexic Advantage*. I didn't tell her about the book and my discovery about myself because I didn't want her to think I was blaming her for failing to recognize it. After all, my brother, Richard, who died of Crohn's disease in 2009, never exhibited any signs of dyslexia, and it would have been hard to tell with my sister, Nancy Ann, who died in 2017.

Now I'm always on the lookout for others with dyslexia, especially children, and I encourage their parents and teachers to nurture the gifts they have. When I give talks to kids, I'll say at the beginning that I'm dyslexic and ask if any of them are. No one raises a hand. But by the time I finish my presentation, I ask them again, "How many of you are dyslexic?" And now they raise their hands. They have learned that someone just like them has followed his passion and gone on to have a successful life. That is what I want them to remember. Not that I found *Titanic,* but that I set goals and kept working to achieve them—and that my dyslexia actually helped me get to where I am today.

—

RELAXING AT THE JEFFERSON HOTEL after a jam-packed day of talking, listening, thinking, and dreaming with my fellow National Geographic Explorers, we always take stock of what we're working on and what it means for the future. It's also another way to think about the kids I meet, because they're the ones who are going to have to confront some of the most difficult problems we can imagine.

A lot of the explorers I know are turning their attention to conservation. Many of us have spent our careers in far-flung places once rarely seen by outsiders. But now, with the number of people in the world and the advanced technologies of communication and transportation, those places are being seen and visited, and they are becoming more vulnerable to change and exploitation. We know we need to work together to save these places, and we trade ideas about how to protect Earth—and humankind—from threats that are mounting.

Some people look at climate change and the massive spike in global population and assume the planet is doomed. I don't worry about Earth, though. When I looked through *Alvin*'s viewport and saw where fresh lava had recently flowed out from beneath the crustal plates, I could see how Earth keeps regenerating itself, releasing the interior heat it acquired when it was formed and creating new geological formations. That process will continue for billions more years. Earth has survived a lot since its birth 4.5 billion years ago, and it will survive anything we do to it. What I worry about is us.

When I was born, in 1942, Earth's population was 2.4 billion. It is now 7.7 billion, more than a threefold increase, and it is expected to reach 10.5 billion by 2050. Nearly three-fourths of the planet is covered by water, and the land includes the unlivable polar regions and deserts. That means 95 percent of all people live on less than 5 percent of the planet. Earth can only support 10.5 billion people if they're all vegetarians. If all humans ate like Americans, Earth could support only 2.5 billion.

We've seen during the coronavirus pandemic what happens when a problem reaches global proportions. By 2050, we could be running short of food. We're losing millions of acres of farmland each year to build homes and suburbia, and we'll lose more to rising sea levels. We need to expand food production, and I think the answer lies in aquaculture, moving from being hunters at sea to being farmers and herders. We've already caught 90 percent of the big wild fish that lived in the oceans—predators like tuna and salmon—so it's going to take a more careful approach to cultivating the seas if we're going to make this work.

The Chinese, for instance, grow kelp twisted around ropes in the sea. We're going to have to scale up things like that and expand to more crops. I've visited Neil Sims, a marine biologist in Hawaii, to see his operation. He raises yellowtail fish 2,000 at a time in floating pens miles out at sea, feeding them algae and soybean protein—one creative approach to more eco-friendly fisheries. With enterprises like his, we can take the pressure off the wild fish stock, and the world's fisheries could bounce back, too.

The most important image to come back to us from outer space was the picture of this little blue marble in a black velvet void of nothingness. With that photograph, Earth became finite. It became precious. If you say there's another blue marble somewhere else in the universe, I will agree completely. And then I'll ask you to calculate how long it's going to take us to get there. Think hundreds to thousands of years, if we're lucky. Are you ready to go?

During the last Explorers Festival, I was speeding through the cafeteria to grab a quick lunch. I spotted five kids, four girls and a boy, all around eight years old. I slowed down.

"Are you making the world better?" I asked, leaning down to get closer to eye to eye.

I got a "yes" from one, and some shy hidden faces from the others, but I'm satisfied I lodged the question in their brains.

Earth is it, folks. Take care of it. There is no Plan B.

CHAPTER 17

SEARCHING FOR AMELIA

E very time I think I'll sit back and watch my younger colleagues take over, something new comes along, and I just can't say no.

In the summer of 2019, I took off on a journey to solve one of exploration's greatest mysteries: What happened to Amelia Earhart? Intrepid pilot and female icon of the early 20th century, she was trying to circumnavigate the globe when she disappeared near some tiny islands in the Pacific. After *Titanic*, Amelia was the biggest thing out there for an explorer like me. Her disappearance in 1937 had long been one of the most compelling searches, and I finally had a shot to pursue it.

The first woman to fly solo across the Atlantic, Amelia captured the hearts and minds of the American people and illustrated what a true adventurer could be. For both my mother, born in 1917, and my grandmother, born in the late 1800s in Kansas, Amelia was their hero—a bold woman who took on the world as few women did then. I felt an instant kinship with her—and not just because she was also born in Kansas.

Presuming she had crashed in the sea or landed on one of the islands, the U.S. Navy and Coast Guard launched one of the largest search efforts in aviation history. They found nothing. It seemed likely that her plane either crashed into the Pacific Ocean or landed on the small island of Nikumaroro. Other than that, little was known about her strange disappearance.

For years, I had ruled out any mission to find her, given the dearth of evidence and the wide expanse of sea where she might have gone down. But in 2012, I gave a presentation at the Pentagon organized by my friend, Peter Lavoy, the Assistant Secretary of Defense for East Asian and Pacific Affairs. After the briefing, Peter took me aside and said I needed to meet his counterpart in Pacific Affairs. He didn't say why.

I soon found myself talking with Assistant Secretary Kurt M. Campbell, a naval intelligence officer I had worked with years before. He closed his office door, handed me an old black-and-white photo, and in a hushed tone asked, "What do you see?"

I scrutinized the image.

"Looks like an old shipwreck resting on a reef off a small island," I said.

"What else do you see off to the left?"

"Something is sticking out of the water."

He then handed me a second photograph.

"We had that image enhanced," he said, "and our experts say it is the landing gear of a Lockheed Electra L10E, identical to the landing gear of Amelia Earhart's plane." That blur in the photo, potentially identified as a telltale clue, came to be known as the Bevington object, named after the photographer who took the picture in 1940.

"Well," I said. "That does narrow down the search area now, doesn't it?"

Campbell asked if I would join Secretary of State Hillary Clinton in endorsing a search effort in the area. It would be led by Ric Gillespie, who heads The International Group for Historic Aircraft

Recovery (TIGHAR) and had spent his lifetime searching for Amelia's plane. We were coming up on the 75th anniversary of Amelia's disappearance, and the secretary of state, apparently a longtime Amelia fan, thought it was time to reopen the search.

So began my immersion in all things Amelia.

Amelia and her navigator, Fred Noonan, began their final flight heading east from Oakland, California, in May 1937. By the time they reached New Guinea nearly a month later, they had flown roughly 22,000 miles, and Amelia had reported plane troubles, weather issues, and personnel tensions, which many believe referred to Fred's drinking problem. The last stretch of 7,000 miles back to Oakland would take her over mostly empty ocean. She and Fred aimed to refuel at Howland Island, a tiny, uninhabited speck in the central Pacific. As they neared the island, low on fuel, Amelia and Fred encountered overcast skies and radio transmission problems. They couldn't find Howland, and a Coast Guard ship sent there to help them land and refuel couldn't track them.

At the same time, a teenage girl named Betty Klenck Brown, living in St. Petersburg, Florida, was operating a shortwave radio with a larger-than-normal antenna that made it possible for her to pick up distant calls, and was writing what she heard in her notebook. One afternoon she heard a voice say, "This is Amelia Earhart," and for the next three hours she recorded messages that seemed to show that Amelia had landed her plane on a deserted island and that Fred was with her, apparently injured.

Betty took down numerous other notes, which led me to believe she was on the right track. Over the years, some explorers have singled out Nikumaroro (also known as Gardner Island), 350 nautical miles south-southeast of Howland, as the most likely place for Earhart to have landed, and the island has been combed for clues. The shipwreck I saw in Kurt Campbell's photo was a steam freighter, S.S. *Norwich City*, that had run aground in a storm in

1929. TIGHAR enthusiasts and others combed the island over and over, and they had found intriguing objects, including a glass jar of what seemed to be freckle cream and pieces of glass from a mirror compact with flecks of dried makeup nearby.

With all this evidence, I endorsed Gillespie's 2012 trip to Niku-maroro, but his expedition failed to find conclusive evidence of her plane. His group only had five days to search, and lost time when one of the vehicles got stuck. His equipment broke down, he had underwater terrain navigation problems, and his search technology just didn't cut it.

After that, I pushed Amelia to the back of my mind. It wasn't until I brought *Nautilus* to the Pacific in 2015 to document the EEZ that the mystery of Amelia's disappearance beckoned again.

Nautilus was scheduled to map the terrain near American Samoa and north to Howland Island—U.S. territory, hence part of the U.S. EEZ—in the summer of 2019. To get there, the ship would pass right by Nikumaroro. But because Nikumaroro is not in U.S. waters, I had to find another way to pay for any sort of detour. It was time to go to National Geographic, my tried-and-true explorer patron.

But the question remained: Could I do it? After all the wrecks I've located over the years, everyone expects me to be the guy who can find the needle in the haystack, but this time, I had a lot less to go on. With *Titanic, Bismarck,* and *Yorktown,* I knew the wrecks existed and approximately where they had gone down. But with Amelia and her plane, evidence was scarce. We really didn't know what had happened when she disappeared. All I had to go on was a grainy photograph and a teenager's diary.

I didn't have a dog in this fight. Scientists should never have a vested interest in the answer; they should just want to know the truth. As I sat in my home office in Connecticut, amid world maps and expedition memorabilia, I knew I had to think everything through carefully. National Geographic had given me the green light for a search of about 20 days and would be producing a television special on it. I had to come up with a plan to give my search an edge that all the other explorations had lacked.

I needed my A-team, including Allison Fundis, my right-hand tactician and chief operating officer at the Ocean Exploration Trust. Allison, who was 38 and has a master's in marine geology, had been with me for more than six years, and she knew how to execute my vision.

My daughter, Emily, also came on board. She had just earned her degree from Ithaca College's Roy H. Park School of Communications with a passion to become a television producer like her mother, and she had worked in video production on several of my cruises. Barbara was also planning to come, but she took a bad fall from a horse and had to stay home rehabbing her leg. I was so glad to have both Allison and Emily on the expedition, along with a host of other women, as we searched for a great female explorer.

Chris Weber, who had produced many of my television shows for National Geographic, became my go-between with Ric Gillespie and TIGHAR. She had worked with them earlier, developing a show on Amelia for Discovery, and now she was helping me navigate our own approach, based on what they had learned. I thought some of their expeditions were amateurish, but I couldn't get caught up in competitive emotions. I needed to focus on the cold, hard facts.

Two theories dominate. There is the Ran Out of Gas and Crashed into the Sea camp, and the Landed on Nikumaroro Island camp. The first posits that when Earhart failed to find Howland Island, she flew back and forth in a north-south direction until she ran out of fuel and crashed into the sea. The second hypothesizes that she landed on a coral reef off Nikumaroro and survived briefly on the island while the tides swept her plane into the ocean, where it sank.

Those who believed the plane crashed into the sea had conducted three major searches off Howland Island between 2002 and 2009. But they had come up with no new clues.

It seemed likely that she missed Howland Island, because the Coast Guard ship waiting for her there reported hearing her signal grow stronger and stronger, then weaker and weaker. If she overshot Howland, landing on Nikumaroro seemed the most likely course of action.

Nikumaroro presented a smaller search area, which I liked. The Nikumaroro theory also had some compelling supporting evidence, including her line of position, as radioed to a Coast Guard cutter on July 2, 1937. Pan American World Airways radio stations had tuned in to her unique frequency, which seemed to be coming from a fixed spot on Nikumaroro. Betty's notebook pointed to an island location, and then there was the Bevington object, the potential Rosetta stone. Remains had been found back in the 1940s on the island, too, although no one knew if they were human remains, let alone those of Earhart or Noonan. Then there were the artifacts that might have belonged to Amelia. Still, it was a gamble because all the other search parties had failed.

I leaned back and gazed out of the window, putting myself inside the cockpit with Amelia. She's flying south, say, having missed Howland Island. She knows she's running low on gas, and she's looking for another island. She sees the *Norwich City* shipwreck, crosses over the island, and looks for a place to land.

If she can see the ship on the reef, it must be low tide, with flat ground running along the surf line. She banks her plane and begins her descent, passing over the shipwreck and touching down to the north. Her wheels make contact with the island's rocky beach, which tears off one of the landing gears. Her plane continues until it comes to rest a short distance north of the Bevington object, near the outer edge of the reef. She and Fred set up camp on the island, and she returns to the plane over the next four to five days at low tide to start one of the engines to operate her radio, perhaps sending the messages that young Betty Brown picked up on her shortwave radio in Florida.

Though Gillespie's team had searched for the plane or its broken-up remains to the north of the possible landing gear without success, I thought I would have a better chance. It was the most logical place to look, and I had an incredible team of engineers and technicians and the most advanced search technologies in the world. Either we would find her plane, or we would rule out the Nikumaroro hypothesis. That is what science is all about: testing hypotheses and ruling out the ones that don't stand up to scrutiny.

But should I temper expectations, and remind people that we were simply testing a theory? Or should I express confidence that we would find her?

We announced the mission at the Television Critics Association convention in Los Angeles. A blitz of interviews followed. As I explained my plan, the instinctive salesmanship and bravado that has helped me raise money and awareness for so many expeditions took over. Finding Amelia was the last frontier for an explorer like me, I told the reporters. I gave them nothing but energetic declarations promising success.

"Amelia Earhart has been on my sonar screen for a long, long time. And I've passed on it," I told the *Washington Post*. "I'm in the business of finding things. I don't want to not find things."

"I wouldn't be going if I wasn't confident," I told the *National Post*, a Canadian newspaper. "Failure is not an option in our business."

We made a promotional video for the Nat Geo show, and I boldly declared: "If this plane exists, I'm going to find it."

But privately, I told Barbara that either I would find Amelia's plane in the first 48 hours of searching, or I wouldn't find it at all.

—

WE ARRIVED AT THE SITE on August 8, 2019, steaming in on *Nautilus* from American Samoa. An idyllic part of Micronesia, Nikumaroro is an isolated coral atoll surrounded by turquoise water turning quickly to deep ocean blue, the island carpeted in a verdant expanse of tropical greenery. We surveyed the island with camera-carrying drones provided by National Geographic. The video they shot was stunning, revealing a thin ribbon of white-sand beaches encircling the vegetation and buffered by shallow waters that splashed over the reefs. Birds soared overhead, and those in our party lucky enough to wander onshore found the island filled with animals chattering in the brush. Allison said she imagined Amelia landing there, becoming a castaway, and then perishing in this tropical paradise. It was a sad and humbling thought. If this was indeed where Amelia and Fred had crashed, at least they died surrounded by beauty.

But we weren't there to admire the scenery. Frankly, I was more interested in what we would find beneath those picturesque waves. I'd always brought great toys on my expeditions, but nothing like the technological extravaganza we'd gathered to look for Amelia. In addition to *Nautilus*'s hull-mounted multibeam sonar mapping system, we had a suite of drones, my trusty ROVs *Hercules* and *Argus,* and *Ben,* the autonomous surface vessel we'd been using to map the caves along the California coast. An onboard computer controlled *Ben,* mapping and photographing the shallower waters around Nikumaroro, while *Nautilus* focused on the deeper areas.

I calculated that Amelia's plane might have come to rest about 1,300 to 1,400 feet northwest of the wreckage of *Norwich City,* on the same coral reef where the Bevington object appeared in the photograph. Low tide could have given her enough solid surface for landing, and even if she had had to land mid-tide, the empty fuel tanks might have kept the plane afloat until they lost their air pockets. That would mean the plane sank farther offshore.

Nikumaroro is a dormant volcano some 4.7 miles long and 1.6 miles wide, and the terrain underwater continues that same downhill slope, ranging from vertical to 45 degrees. That means that anything falling off the reef would slide down and probably splinter along the way, as if it were tumbling down a stone staircase with lots of steep, hard falls. So we'd be searching for her plane—or, more likely, its fragmented remains—on the rugged side of the volcano, which was already strewn with large blocks of coral that had been tumbling down the slope for thousands of years. The slope would be streaked with small gullies and ridges, too, making it even trickier to find plane fragments among the rocks and sand. We would search this rocky surface, beginning in deep water and gradually working our way uphill. Just to be safe, we'd also check south of the *Norwich City* and the shallow slopes around the entire island.

First we constructed what I called my battle map—a detailed three-dimensional map made by combining coordinated sonar readings from *Ben* and *Nautilus.*

Nikumaroro is such a small slip of land that in less than a day, we were able to circle it five times with *Nautilus,* each time scanning a new depth, down to more than 12,000 feet. *Ben* moved more slowly, with a narrower scan width, and went around the island three times. We documented the movement of debris from *Norwich City,* knowledge that could help us search for the plane's debris to the north.

Drones circled the island and collected 3,000 images. Water clarity was so good, we could see down to 65 feet, and we spent hours poring over those images, looking for pieces of a plane. *Hercules* mostly launched at night, its high-definition video camera working in areas illuminated by *Argus*'s floodlights. Our divers did practice dives and were ready to retrieve anything we found as soon as we gave the word.

We kept a 24-hour vigil in the dimly lit control room of *Nautilus,* trading off in four-hour shifts to make sure we always had eyes on the search area. Members of my team kept their headsets on, donning fleeces as they came into the control room, which we kept freezing to protect the electronics. Outside, the air was hot and humid—so hot that a team that went ashore with bone-sniffing dogs had to limit their work because the ground scorched the dogs' feet.

Sitting in the icy dark of the control room, we rarely spoke. We just murmured comments, our senses heightened, our eyes transfixed on the digital displays that might reveal a hint of plane debris. Every time my crew spotted something man-made, we had to take a closer look. The ROV pilots would pick up the object with the robot hands and position it for *Hercules'* camera or recover it for closer inspection.

But over and over again, a closer look revealed nothing of interest: a piece of *Norwich City,* or a long-degraded soda can, or some foreign debris lodged in the rubble. We found an old side-scan sonar, an artifact from a previous explorer.

We kept coming up with nothing. As the hours ticked by and the first days passed, I reminded my crew that deepwater searches take persistence. *Titanic* wasn't found on the first go-round, nor was *Bismarck*. We were in it for the long haul. But as I had told Barbara, the first 48 hours were key. About 72 hours in, I knew in my gut we were looking in the wrong spot.

I pulled Allison aside, away from the rest of the crew. She had been planning this search for a year and felt such a personal connection to Amelia.

"I don't think she's here," I said. "But we're not going to tell the team that."

Allison had privately come to the same conclusion. A lot was riding on a blurry photograph, she said.

We had to press on, expand the search area, and think bigger. We both knew that. If the point was to test a hypothesis, we had to pull out all the stops to make sure this one was dead. It became less about finding Amelia, and more about scientific rigor. We were going to test every theory to the limit.

As we learned more about the terrain and the underlying rock outcrops making up the island, we kept coming up with more theories about what could have happened. Each day, we would strategize, revise and review, and determine the best course of action. It was exhausting, but the adrenaline keeps you going.

I was sleeping in short bursts and spending all my time clinging to the monitors in the control van. A lot of my team members were young and had never seen me in my zone, laser-focused on the mission. Even Allison, who had been on seven expeditions with me and supported many more from shore, said she had never seen me so intense. We went through every theory, disproving each one as we scanned the ocean floor. By the end, we had visually inspected 100 percent of the target areas down to a depth of 1,000 feet—3,000 in some more focused areas—and we found no evidence of Amelia Earhart's plane.

The younger crew members were watching me as the days wore on. I made a point of never letting them see me down—that's a key to leadership I learned long ago. I cheered them on and kept pushing them to try new tactics, test new theories. I never shared my doubts, not even with my daughter.

As the days elapsed, they started playing more card games. Emily proved herself particularly good at acey-deucy, a game like backgammon. I joined in some of the games and learned more about some of the younger members of the team. I learned that one of them, Megan Lubetkin, has dyslexia, and she caught my attention with an idea to teach people about science through the visual arts.

By the end of the expedition, I had pulled out every tool I had in my search arsenal, and we proved that Amelia simply wasn't there. We all know we gave it a good search. It's not the outcome that we'd hoped for, but you cannot force her plane to be where it's not.

August 21 was our last full day of operations. We had hours of video for the television special, and it was a good story, no matter how it ended. As with *Bismarck* in 1988, I could say one thing for certain: I knew where Amelia Earhart wasn't.

—

ALTHOUGH WE DIDN'T FIND AMELIA, I feel humbled at having been part of the search effort around her disappearance, and thankful for the chance to bring some exceptional people with me along the way. The search for a great woman and the relics of her success felt personal, important, and timely. It's taken too long, but we're doing a better job of encouraging women to be scientists, engineers, and leaders. I tell Emily and the other young women working with me not to let anyone stop them from pursuing their dream, no matter how crazy it might sound. What could be crazier than wanting to be Captain Nemo? Having Allison co-lead this expedition and all the women in our Corps of Exploration on this trip meant the world to me and felt like a fitting tribute to Amelia's memory— so many female explorers, united across the decades. That's what brings me the most pride in this mission, despite the outcome.

In an explorer's line of work, there is always something at risk. Whether it's your reputation or your very life, if you don't put what you find most valuable on the line, the search loses its soul. Going into the fray, you have to be confident. If you don't quit, you're not out yet.

I haven't quit. I'll be back.

National Geographic wants to return in 2021 to search the other side of Nikumaroro for any signs that she might have been there. If that fails, we will set our sights on the Howland Island theory in 2022. The vastness of that search area presents a host of problems. *Nautilus* started mapping the underwater terrain around Howland Island shortly after leaving Nikumaroro in 2019. We needed to do that as part of our exploration of America's extended exclusive economic zone for NOAA, and building that map will help lay the groundwork to possibly deploy a swarm of autonomous undersea vehicles that could search that vast area together like a pack of wolves and help us compensate for the lack of specific clues.

As I realized when I got rejected from Scripps when I was all of 20-something, every failure is a learning lesson. Since then, I've grown to love failure. It's not something you should try to avoid, but rather embrace and learn from—and then beat. You don't go around it—you go through it. You get knocked down, and you lie there, and you go, *Wow, that was a hell of a shot.* And then you dive back in. I've failed lots of times, but I don't quit, so I always win.

We're all on a journey. The explorer in me loves the challenge, the chase, the search, the raw energy, and the rush—the thrill of finding something you never set out to find, as well as the sweet victory of discovering what you were looking for all along.

So I'm here to say, I ain't giving up. I'm waiting for newer technologies, a brighter day, a calmer sea. It might not be me who finds Amelia. It might be Allison, or Emily, or someone else in a coming generation. Or maybe Amelia will never be found, yet all we learned searching for her will lead to some other discovery.

You have to dream big to make a difference in this life, and I intend to keep on dreaming. A world of discovery still awaits.

EPILOGUE

Even now, as I approach my 80s, I find myself guided and inspired by Joseph Campbell's concept that life is a journey, an act of becoming, and that you never really arrive at your destination. I thought about that concept again recently, as I lay in the grass under a grove of redwood trees with my son Ben, listening to Arthur Brooks, a Harvard professor, give a talk about the secrets of getting happier as you get older.

As Brooks got around to talking about people 75 years and older, my ears perked up. At this point in life, he said, there are three things you need to do.

The first is to develop deeper friendships, and he had the perfect analogy for it. He urged us to look at the giant redwoods. They can grow to more than 300 feet tall, even though their roots sink less than six feet deep. How do they keep from falling over as they grow older? They reach out to the other trees around. Like the redwoods, Brooks suggested, we must develop a deeper network of friends and family, intertwining our roots so we stand tall together.

For me, that's easy. When I'm not in my garden or greenhouse, I'm spending more time with friends, whether relaxing at dinner parties with Barbara and other couples or fishing with my buddies in the local streams or out in the salt water just minutes from my home. I'm also spending more time with my grandkids, Dougie's sons, Carter and Blake.

Second, Brooks said, it's important to focus less on yourself and more on mentoring others. I've always mentored my team members, and connecting with the younger generation has long been an important part of my work. So I'm happy to step up that role, especially with gifted young associates like Allison Fundis, who helped me run the Amelia search, and Megan Lubetkin, whose ideas for combining art and oceanography intrigue me to no end. I believe that every generation stands on the shoulders of the previous ones and sees new horizons its predecessors could not see, and I'm excited to see what the next generation of explorers will do.

The third thing Arthur Brooks said you need to do as you speed past 75 toward 80 and beyond is to "just say no." When new ideas enter your mind or new opportunities come along, resist the urge to jump in.

But saying no is against my very nature. My mind still generates a constant flow of ideas, and I've always said yes the moment an opportunity came along that could turn one of my ideas into reality. I still have things I want to do. Besides searching further for Amelia, I'd love to go back to the Black Sea. And I want to explore the Indian Ocean. In fact, when I told Barbara about Brooks's point number three, she chuckled, saying I'd never be able to say no. Ben compares me to a top, always whirling. I might spin a little more slowly as I approach 80, he says, but I'll still be whirling.

He's right. I'm not ready to just sit back in an easy chair. My dyslexic brain likes to visualize all the angles, so I've been thinking about modifying the "no" so it's not so absolute. Although it's exhilarating to search for *Titanic* or *Bismarck,* or to create a program that transports students to the Galápagos or the Amazon rainforest, it's also stressful having to pull together all the people, technology, and money that's needed. It's especially difficult to deal with the government. I've recently lined up hundreds of millions in federal funds to support my Ocean Exploration Trust, building a consortium of universities to continue the work of developing new robotic vehicles and exploring the deep for years to come. So now I feel able to start handing off some of this work to associates like Allison, who also is OET's chief operating officer. The trust has a wonderful board, led by Chairman Robert Patricelli, an unflappable business leader, and Vice Chairman Larry Mayer, a marine geophysicist at the University of New Hampshire who's been a partner of mine for years.

I might still go to sea on occasion, but more as a mentor than the guy directing every step. During the COVID-19 crisis, I've created a communications center in the basement of the old stone house that serves as OET's headquarters. Now I can give lectures and participate in meetings without traveling. I can also follow *Nautilus*'s adventures on the monitors and give advice to Allison and the others aboard without leaving my home base. In many ways, this brings my idea of scientific telepresence full circle, and it's helping me shift from being Warrior Bob, out in the midst of the action, to Virtual Bob, the dreamer and consulting avatar.

I've also hired Megan to work with Emily and me to create DEEP SEA Space, shorthand for Deep Experimental Exploration Program in Science, Engineering, and Art, in cooperation with the Lyman Allyn Art Museum in New London, Connecticut—a program of exhibits and activities to promote ocean awareness and deep-sea curiosity for all ages through a multisensory approach, appealing to the general public and especially kids with dyslexia.

The fact is, I've spent my life dreaming up new ideas, and some of them took 15 to 20 years to materialize. I know I don't have 15 to 20 more years, but I'm not going to stop dreaming. So I've started writing science fiction, conjuring what the world might look like in 2050 and beyond, and envisioning what today's children and their children can do to make it better. Remember, Jules Verne's fictional creation, Captain Nemo, inspired my passion for ocean exploration. Now it feels like it's time for me to be more like Nemo's creator than Nemo himself.

So what have I learned, through both my explorations at sea and my efforts to understand my own special gifts? Trust my daughter, Emily, to find a way to sum it all up. When she came home from college last spring, I noticed she unpacked a little sign with a quotation from J.R.R. Tolkien: "Not all those who wander are lost." Those words of wisdom capture the essence of what explorers do and the spirit that continues to drive my life today. It's a message of hope, confidence, and optimism—a message I want to share with people, young or old, with or without dyslexia, as they continue on their own journeys of discovery.

EXPEDITIONS, PUBLICATIONS & MEDIA

AT-SEA EXPERIENCE

157 Expeditions

- 63 different research platforms
- 3 nuclear submarines
- 10 different deep submersibles and bathyscaphes

ARTICLES IN *NATIONAL GEOGRAPHIC* MAGAZINE

"Dive Into the Great Rift," May 1975

"Window on Earth's Interior," August 1976

"Oases of Life in the Cold Abyss," with John B. Corliss, October 1977

"Return to Oases of the Deep," with J. Frederick Grassle, November 1979

"*NR-1,* the Navy's Inner-Space Shuttle," April 1985

"How We Found *Titanic,*" with Jean-Louis Michel, December 1985

"A Long Last Look at *Titanic,*" December 1986

"Epilogue for *Titanic,*" October 1987

"The *Bismarck* Found," November 1989

"Riddle of the *Lusitania,*" April 1994

"High-tech Search for Roman Shipwrecks," April 1998

"Ghosts and Survivors Return to the Battle of Midway," by Thomas B. Allen, April 1999

"Deep Black Sea," May 2001

"The Search for *PT-109,*" December 2002

"Ancient Wrecks Await: Ballard Expedition 2003," by Peter de Jonge, May 2004
"Why Is *Titanic* Vanishing?" December 2004
"Mapping Offshore America," November 2013

Plus more than 100 scholarly articles in publications including *Science, Nature, Geological Society of America Bulletin, Journal of Geophysical Research, Earth and Planetary Science Letters, Oceanus,* and *Deep Sea Research.*

FILMS & TELEVISION SPECIALS

Exploring the Ocean Bottom: The Cayman Trough, 1976
Dive to the Edge of Creation, 1980
Born of Fire, 1983
Secrets of the Titanic, 1986
Search for Battleship Bismarck, 1992
The Lost Fleet of Guadalcanal, 1993
The Last Voyage of the Lusitania, 1994
The Living Sea, 1995
Titanic*'s Lost Sister,* 1997
The Lost Treasures of the Deep, 1999
The Battle of Midway, 1999
Lost Ships of the Mediterranean, 2000
Lost Liners, 2000
Pearl Harbor: Legacy of Attack, 2001
In Search of Noah's Flood, 2001
Beneath the Sea, 2002
The Search for Kennedy's PT-109, 2002
Lost Subs: Disaster at Sea, 2003
Ghosts of the Baltic Sea, 2003
Quest for the Phoenicians, 2004
Ghosts of the Black Sea, 2008
Titanic: *The Final Secret,* 2008
Deep Secrets: The Ballard Gallipoli Expedition, 2010
Gallipoli's Deep Secrets, 2010
Save the Titanic *with Bob Ballard,* 2012
Alien Deep (five episodes), 2012
Caribbean's Deadly Underworld, 2014
Nazi Attack on America, 2015
Expedition Amelia, 2019

BOOKS

<u>ADULT NONFICTION</u>

Photographic Atlas of the Mid-Atlantic Ridge Rift Valley (1977)

Exploring Our Living Planet (1983)

The Discovery of the Titanic (1987), #1 Best Seller, *New York Times* & *London Times*

The Discovery of the Bismarck (1990), #1 Best Seller, *New York Times* & *London Times*

The Lost Ships of Guadalcanal (1993)

Explorations: My Quest for Adventure and Discovery Under the Sea (1995)

Exploring the Lusitania (1995)

Lost Liners (1997)

Return to Midway (1999)

The Water's Edge (1999)

The Eternal Darkness (2000)

Adventures in Ocean Exploration (2001)

Graveyards of the Pacific (2001)

Mysteries of the Ancient Seafarers (2004)

Return to Titanic (2004)

Collision with History: The Search for John F. Kennedy's PT-109 (2002)

The Lost Ships of Robert Ballard (2005)

Archaeological Oceanography (2008)

Titanic: *The Last Great Images* (2008)

<u>CHILDREN'S NONFICTION</u>

The Lost Wreck of the Isis (1990)

Exploring the Bismarck (1991)

Explorer: A Pop-Up Book (1992)

Exploring the Titanic (1998)

Ghost Liners (1998)

<u>ADULT FICTION</u>

Bright Shark (1992)

ACKNOWLEDGMENTS

First, I want to express my love and thanks to all the members of my family, namely Chester Patrick (my father), Harriett Nell (my mother), Richard Lee (my brother) and his children, and Nancy Ann (my sister); also to Todd Alan, Douglas Matthew, and William Benjamin Aymar (my sons); Emily Rose Penrhyn (my daughter); my two grandsons, Carter and Blake; and Mabel May (my grandmother). But this would not be complete if I didn't thank my wife, Barbara Earle Ballard, and her family, the Earles, who embraced Dougie and me when I had no family close by.

Next are the members of my personal staff who, over the years, knew my every move and, in fact, made sure I made those moves: in particular, Cathy Offinger and Janice Meagher, as well as Angela Souza Keir, Linda Lucier, Gretchen McManamin, Sharon Hunt, Larry Flick, Janet Allen, and Terry Nielson while at Woods Hole, followed by Laurie Bradt, Denise Armstrong, Sandra Witten, and Richard Darling, first at the Institute for Exploration and more recently at the Ocean Exploration Trust.

Over the years I have been lucky to have someone at every step of the way whom I deeply respected and, believe it or not, listened to—people who saw something in this "white tornado from Kansas," who took me under their wing and opened doors I could not open myself. Thanks to Dr. Andy Rechnitzer, Dr. Dick Terry, Dr. Bob Norris, Dr. Kenneth O. Emery, Dr. Elazar Uchupi, Dr. Robert McMaster, William "Bill" Rainnie,

Dr. Robert W. Morse, Dr. Jerry van Andel, Dr. Paul Fye, Dr. Robert Spindel, and Charlie Black.

My life has taken me into harm's way on many occasions, which has required me to work on an open deck at night, to struggle in a gale to recover our vehicles, or to venture thousands of feet beneath the sea to explore a world of eternal darkness and then come back home alive. I want to thank the people who were at my side along the way: Allison Fundis, Earl Young, Steve Gegg, Stu Harris, Tom Crook, John Porteous, Mark DeRoche, Dwight Coleman, Jim Newman, Nicole Raineault, Mike Durbin, Emory Kristof, Katy Croff Bell, Peter Schnall, Scott Monroe, Jonathan Wickham, Chris Weber, Chad Cohen, Martin Bowen, the crews of E/V *Nautilus* and the 60 other ships I have sailed on, my *Angus* and *Argo/Jason* teams, the team at the JASON Foundation for Education, the Alvin Group, and now the Corps of Exploration.

What I do cannot be done alone. It requires collaborators who either share my dream or want it to succeed, including people like Dr. Jean Francheteau, Dr. Larry Mayer, Dr. Dana Yoerger, Tufan Turanli, Jean-Louis Michel, Dr. Jerry Wellington, Dr. Larry Stager, Dr. Fred Hiebert, Dr. David Mindell, Dr. Jamie Austin, Jacquie Hollister, Dr. Art House, Dr. Peter Girguis, Andy Bowen, Skip Marquet, Mike Stewart, Dr. Steve Carey, Dr. Haraldur Sigurdsson, Dr. Michael Brennan, Dr. Bruce Corliss, Capt. Jack Maurer, Dr. Hagen Schempf, Tim Delaney, Dr. James Delgado, Cmdr. Jason Fahy, Megan Lubetkin, Andy Fedynsky, Mike Gustafson, David Weil, Mary Norris, Ken Marschall, Dr. Enric Sala, Kristin Rechberger, Phil Segal, Tim Armour, and Barrie Walden.

Then there are those who "bet on my horse," who made it possible for me to do what I do, leaders in the multifaceted worlds I live in, including those in the private sector who are risktakers like Robert Particelli, Riley Bechtel, Vinnie Viola, David O'Reilly, Charlie Johnson, Wayne and Marti Huizenga, Mike Bonsignore, Don Koll, Marco Vitulli, Harry Gray, Wes Howell, Steve Wynn, Jed Hammer, and William E. Simon, along with key members of our state and federal governments, including Senator and Governor Lowell Weicker, Governor Don Carcieri, University of Rhode Island President Dr. Robert Carothers, Congressman John Culberson,

Senator Fritz Hollings, and Secretary of the Navy John Lehman, Jr.; in the Office of Naval Research, Dr. Frank Herr, Dr. Fred Saalfield, Dr. Gene Silva, and Dr. Tom Drake; at the Department of Commerce, Craig McLean and Dr. Alan Leonardi; as well as those in uniform, including Admirals Dwaine O. Griffith, Ron Thunman, Jonathan W. Greenert, Edmund P. Giambastiani, Jr., Jay Cohen, Paul Gaffney, and Tim Gallaudet. And then there were Hugh Connell, Howard Stirn, and Dr. Jerry Burrow, who welcomed me with open arms when I left Woods Hole and came to the Mystic Aquarium/Sea Research Foundation.

In addition to those already mentioned, many at National Geographic, which I have called home since 1974, have supported me throughout my career. They include John Fahey, Gary Knell, Mike Ulica, Dr. Vicki Phillips, Gil Grosvenor, Terry Garcia, Chris Liedel, Tim Kelly, Terry Anderson, Sam Matthews, Bill Garrett, and Alex Moen, as well as those who made this book possible: Lisa Thomas, Susan Hitchcock, Hilary Black, and Adrian Coakley, who worked so hard with my two principal collaborators, Chris Drew and his wife, Annette. — R.D.B.

—

I couldn't have written this book with Bob without the help of my amazing wife, Annette Lawrence Drew. She has a Ph.D. from Princeton, and she once again put her research skills to wonderful use. She participated in most of the interviews with Bob and organized a huge amount of material on his early years, his dealings with the Navy, the *Titanic* expeditions, and his discovery of the book *The Dyslexic Advantage,* which changed his life. She got so deeply into her work that she surprised Bob with questions on things he'd never considered about the *Titanic,* and he couldn't believe how well she understood him after she put together a section on how the Eides' findings on dyslexia applied to him. Annette also kept us all going, as did the support we received from our daughter, Celia; my always cheerful 90-year-old dad, Leon; our siblings Mark, John, Mary-Elise, Betsy, Laura, and Cindy, and spouses Priscilla, Catherine, and David; and close friends Curt and Sharon Hearn. I love and appreciate all of you.

The deadline for this book was originally going to be very tight, and Bob has lived an epic life, so I assembled an all-star team of journalists and journalism students to help us with the research and writing. Rick Lyman, a correspondent and editor retired from the *New York Times;* John Diamond, a former Associated Press intelligence reporter; Zachary Fryer-Biggs, a national security reporter for the Center for Public Integrity; and Jacqueline DeRobertis, my former graduate assistant at LSU who is now a reporter for *The Advocate* in Baton Rouge all wrote excellent drafts of chapters. Current and former LSU communications students Kaylee Poche, Mary Chiappetta, and Rachel Mipro transcribed interview tapes and organized the material. I knew everyone on our team was doing exceptional work, and I really got to savor it after I went back through everything to make the finishing touches. Martin Johnson, the late dean of LSU's Manship School of Mass Communication, and the LSU Foundation also provided great support to me and some of the researchers.

Bob asked Lisa Thomas, publisher and editorial director of National Geographic Books, and Hilary Black, her deputy editor, to see if I would write this book with him because he loved *Blind Man's Bluff,* a best-selling look at submarine spying during the Cold War that Annette and I wrote with our friend Sherry Sontag. That book led us to Esther Newberg, the best literary agent in New York, who is fiercely loyal to her clients and a caring friend who helped make sure this project came together. And a special thanks to Susan Hitchcock, a senior editor at National Geographic Books whose strong sense of narrative—and appreciation for our efforts to capture Bob's distinctive voice—helped us tell what we think is a riveting tale. — C.D.

SOURCE NOTES

Most of this memoir—a much more personal recounting of my journey and all my expeditions than I have ever given—comes from memory, as captured in nearly 200 hours of interviews with my co-author and his wife, Annette. I like to say that they put me on the couch and helped me understand myself and my experiences better than I ever had before. I also worked with great colleagues to create books and television specials after each of my major expeditions, and I occasionally consulted them to refresh my memory as we worked on this book. We also consulted other sources for additional material, and what follows is a list of those sources for each chapter.

CHAPTER 1: FINDING NEMO

My father, Chet Ballard, wrote an unpublished manuscript, *Raisin' Richard, Robert, and Nancy*, about how he and my mother, Harriett, raised their three children. It was filled with details, and I was grateful to have it to refresh my memory.

My mother, Harriett Ballard, kept every article, report card, and accolade about her children. I also used her scrapbooks to flesh out my early years. I took after my mother in saving all my own articles, correspondence, phone logs, and personal and professional communications throughout my career. These files also were invaluable in putting this memoir together.

CHAPTER 2: SWISS ARMY KNIFE

Pages 34–35: In describing Fred Spiess, we consulted his obituary notice on the website of the Scripps Institution of Oceanography: *scripps.ucsd.edu/news/obituary-notice -pioneer-ocean-technology-fred-n-spiess*. Additional information comes from an interview for Dr. Spiess's oral history that Christopher Henke conducted for the University of California, San Diego, in March 2000.

CHAPTER 3: BECOMING A SCIENTIST

Pages 44–45: Some of the information about the Sea Rovers comes from *The History of the Boston Sea Rovers,* a manuscript that the group prepared. I have a preview copy of it.

Page 45: Joe Hohmann's quote referring to my interest in finding *Titanic* as "a pipe dream" comes from a letter from Joe included in *The History of the Boston Sea Rovers,* page 94.

Page 57: Basic information about all of the dives that *Alvin* has made is available in the *Alvin* dive logs on the website of the Woods Hole Oceanographic Institution: *dsg.whoi .edu/divelog.nsf.*

CHAPTER 4: REWRITING THE SCIENCE BOOKS

Page 61: Research science is very much a team endeavor, and I was one of seven authors of the article in *Nature* that I mentioned when my snotty colleague questioned why I was writing for *National Geographic* magazine. The *Nature* article was titled "Inner Floor of the Rift Valley: First Submersible Study," and it appeared in 1974 in volume 250, number 5467, pages 558–60. Interestingly, although many scientists once looked down on colleagues who explained their work in the popular press and on television, the National Science Foundation now requires that grant applicants create plans to make the public aware of their research findings.

Pages 76–77: These additional articles in *Nature* and *Science,* two of the most prestigious peer-reviewed journals, described the findings about hydrothermal vents and black smokers. Once again, I shared credit on the articles with the other amazing research scientists on these expeditions. The article that won the Newcomb Cleveland Prize as the best article in *Science* in 1980 was titled "Hot Springs and Geophysical Experiments."

CHAPTER 5: MAKING FRIENDS WITH THE NAVY

Pages 79–80: Our description of *Titanic*'s last hours is partly based on accounts from Walter Lord's masterpiece *A Night to Remember,* published in 1955. Lord located 63 survivors to help him put together a detailed and harrowing account of what happened on that tragic April night in 1912.

Page 86: Jack Grimm's interest in using a monkey to point to *Titanic*'s location on a map comes from Jack Grimm and William Hoffman's book, *Beyond Reach: The Search for the Titanic,* published by Beaufort Books in 1982.

Page 90: In an article in *Navy Times* on April 8, 2013, the 50th anniversary of *Thresher*'s loss, Bruce Rule and Norman Polmar gave a concise account of *Thresher*'s demise that went beyond the initial findings the Navy made public in the 1960s. Rule had helped run the Navy's underwater sound surveillance system (SOSUS), and he and Polmar wrote that acoustic evidence indicated that an electrical bus failure caused the reactor's coolant pumps to shut down. This, in turn, led to a shutdown of the nuclear reactor (a "scram") at the sub's test, or maximum operating, depth of 1,300 feet. Once a nuclear reactor scrammed, it could not be restarted quickly, and *Thresher* was too deep to have the luxury of time. As *Thresher* attempted to blow high-pressure air into her ballast tanks to release water and regain buoyancy, ice formed in the lines and collected in strainers, preventing air from pushing out the water in the ballast tanks. Rule and Polmar said the acoustical evidence showed that *Thresher* sank past her presumed crush depth of 1,950 feet (one and a half times test depth) and finally imploded at 2,400 feet. Her steel hull had held for more than 400 feet beyond its presumed crush depth, but *Thresher* was operating in 8,400 feet of water.

When I photographed and mapped *Thresher*'s debris field in 1984, Navy officials told me that they believed a mist of water from the pipe had contributed to the electrical failure. I was witness to how the immense pressure had shredded the submarine. The 129 men on board wouldn't have felt the implosion; it happens too quickly. But they would have known what was coming.

Pages 91–93: Adm. Ronald Thunman and former Navy Secretary John Lehman supplemented my memories about how I lobbied them to search for *Titanic* in interviews with the Ronald Reagan Presidential Library and Museum. The National Geographic Museum used excerpts from the interviews in an exhibit titled "*Titanic:* The Untold Story." Mark DePue conducted Admiral Thunman's interview on March 20, 2017, and Secretary Lehman and I were interviewed together. Lehman provided his recollections on how President Ronald Reagan responded to my plan to find *Titanic* in a conversation conducted for this book in 2019.

CHAPTER 6: MAY GOD BLESS THESE FOUND SOULS

Pages 100–101: Many of the historical details of *Titanic*'s last night come from Walter Lord's book, *A Night to Remember*. Additional historical information can be found in my own book on *Titanic*, *The Discovery of the Titanic*, which chronicles my 1985 and 1986 expeditions. Walter Lord's quote, "We wonder what we would do," and his description of the impending danger as the tragedy unfolded comes from the Introduction to *The Discovery of the Titanic*.

Page 100: In 1940 Jack Thayer, who had witnessed *Titanic*'s sinking as a 17-year-old passenger, self-published his recollections for his family and close friends. His book, *The Sinking of the SS Titanic April 14–15, 1912*, was reissued in a paperback edition by Spitfire Publishers Ltd. in 2019. The quote we use from him is from the paperback edition, page 19.

Page 101: When Jack Thayer was rescued by the *Carpathia*, he described to an artist traveling on that ship what he had seen as *Titanic* sank. These famous sketches of the ship breaking in half are what I recalled when I thought it might be more useful to search for a debris field rather than the ship itself. See pages 28–33 in Thayer's book for facsimiles of the sketches. Thayer was in the water with his life vest, mesmerized by *Titanic* as she sank. Most other survivors described the ship sinking intact, and based on their accounts, Walter Lord believed she went down in one piece. I thought Thayer's account seemed credible because he was so close to the ship itself. In fact, Thayer wrote (pages 38–39) that as the second funnel collapsed, it fell only 20 to 30 feet away from him. He was that close.

Page 111: Jack Thayer's poignant quote about the wailing of drowning passengers sounding like locusts comes from page 41 of his book.

Page 118: The number of photographic frames of *Titanic* comes from my and Jean-Louis Michel's *National Geographic* article, "How We Found *Titanic*" (December 1985, page 717).

CHAPTER 7: TAKING STOCK AFTER *TITANIC*

Page 119: The quote "I'm glad it's over" comes from the *New York Times*, "Airhorns Blare as Titanic Researchers Sail In," on September 10, 1985. The story was written by *Times* science writer William J. Broad, who has done an excellent job of covering many of my expeditions and other ocean exploration.

Pages 129–130: I used details from *Atlantis II*'s status reports to Woods Hole about *Alvin*'s dives at the *Titanic* site. Copies of these status reports are in my personal archives.

CHAPTER 8: MY CINDERELLA STORY

Page 155: The tape of Todd Ballard's interview with National Geographic TV was made available by Chris Weber, who has produced a number of my television specials.

CHAPTER 9: LAID LOW

Pages 162–163: Joseph Campbell describes his philosophy of the epic journey that we are all on in *The Power of Myth* (Anchor Books, 1991) and in other works.

CHAPTER 10: A NEW PARTNERSHIP

Page 185: The *New York Times* and other newspapers ran stories with varying details about the ammunition that *Lusitania* carried, and several books about the ship's sinking discuss the theories about what triggered the mysterious second explosion.

CHAPTER 11: ROLLING THE DICE

Pages 194–195: *Dark Waters,* by Lee Vyborny and Don Davis (New American Library, 2003) provides an excellent history of the Navy's *NR-1* submarine.

Pages 195–196: David A. Mindell's recollection of the 1997 cruise looking for ancient shipwrecks is recounted in his book, *Our Robots, Ourselves* (Viking, 2015, pages 19–24).

Page 209: We recorded most of the conversations in our ship control rooms for the television shows on our expeditions, and *Yorktown* veteran Bill Surgi's emotional quote about seeing his shipmates doing their jobs again as we found the wreckage comes from those tapes.

CHAPTER 12: BLACK SEA QUEST

Pages 211–212: Some of the details of Willard Bascom's biography come from his obituary in the *Los Angeles Times*. It was written by Myrna Oliver and published on October 14, 2000. John Steinbeck described his travel with Bascom in *Life* magazine's April 14, 1961, issue, pages 111–22. Bascom's book, *Deep Water, Ancient Ships: The Treasure Vault of the Mediterranean,* was published by Doubleday in 1976.

Pages 213: *Noah's Flood: The New Scientific Discoveries About the Event That Changed History,* by William Ryan and Walter Pitman, was published by Simon & Schuster in 1998.

Pages 217–220: The direct quotes from me and Harvard professor Larry Stager on the cruise examining Phoenician wrecks were recorded by a National Geographic Television crew and featured in my show, *Quest for the Phoenicians,* produced by Pamela Caragol.

Page 222: Dale Keiger interviewed George Bass for an article called "The Underwater World of George Bass," published in *Johns Hopkins Magazine* in April 1997. It provides an excellent account of his role as founder of marine archaeology and his major explorations.

CHAPTER 13: CLOSE-UPS ON THE PAST

Pages 227–228: *Endurance: Shackleton's Incredible Voyage,* by Alfred Lansing, provides a hair-raising account of Sir Ernest Shackleton's travails with *Endurance*. We read a paperback edition published by Basic Books in 2014.

Page 241: This article contains links to the history of U.S.N.S. *Capable,* converted to an oceanographic research vessel, *Okeanos Explorer: en.wikipedia.org/wiki/USNS_Capable_(T-AGOS-16)*.

Page 243: The Eva Hart quote is from the *National Geographic* article "How We Found *Titanic*," by me and Jean-Louis Michel, December 1985, page 718.

CHAPTER 14: BECOMING CAPTAIN NEMO

Page 250: The description of the ornate clock in the Waldorf Astoria Hotel is drawn from an article, "The Fantastic Clock in the Lobby of New York City's Waldorf-Astoria Is Pure History," by Elizabeth Doerr. It was published in 2017 and is available at *quillandpad* *.com/2017/02/28/fantastic-clock-lobby-new-york-citys-waldorf-astoria-pure-history*.

CHAPTER 15: ALL THAT *NAUTILUS* COULD DO

Page 276: The description of the German campaign in the Gulf of Mexico was drawn partly from *Operation Drumbeat: The Dramatic True Story of Germany's First U-Boat Attacks Along the American Coast in World War II*, by Michael Gannon. It was published in 1990 by Harper & Row.

Pages 277–279: The well-known shipwreck diver Richie Kohler was with me when we found the wreckage of the German U-boat in the Gulf of Mexico, and our account of it draws in part on an article that Richie wrote for *Wreck Diving Magazine*, a quarterly. The article was called "U-166 and the *Robert E. Lee*," and it appeared in Issue 34.

CHAPTER 16: DISCOVERING MYSELF

Pages 285–287: Brock L. Eide and Fernette F. Eide's book, *The Dyslexic Advantage*, was a game changer for me, literally helping me to discover me. The book was first published in 2011, but I did not hear about it until 2015. It is available in paperback from Plume, a division of Penguin.

Page 288: The 5 to 17 percent estimate of the number of people in the world with dyslexia comes from a University of Michigan website, *dyslexiahelp.umich.edu/answers/faq*, accessed on August 31, 2019.

CHAPTER 17: SEARCHING FOR AMELIA

Pages 295–296: In writing our account of Amelia Earhart's disappearance and the previous searches for her, we consulted *Finding Amelia: The True Story of the Earhart Disappearance*, by Ric Gillespie, the executive director of the International Group for Historic Aircraft Recovery (TIGHAR). It was published by the Naval Institute Press in 2006.

Page 299: The *Washington Post* story, under the headline "This Explorer Found the *Titanic*. His New Mission: Solve Amelia Earhart's Disappearance," was written by Alex Horton and appeared on July 24, 2019: *washingtonpost.com/history/2019/07/24/this-explorer-found-titanic-his-new-mission-solve-amelia-earharts-disappearance*. "'Failure Is Not an Option': Can the Man Who Discovered the *Titanic* Find Amelia Earhart's Plane?" by Bianca Bharti appeared on July 25, 2019, in the *National Post, national post.com/news/world/amelia-earhart-plane-robert-ballard-search-nikumaroro*.

INDEX

Boldface indicates
illustrations.

A

ADHD 288–289
Advanced Tethered Vehicle
(*ATV*) 206, 207–208
AE2 (submarine) 256
Aksoy, Hasan Huseyin 268,
269
Alberthal, Lester, Jr. 144–
145, 164
Alcoa 81–82, 212–213
Alexander von Humboldt
(ship) 247–251
Alvin (submersible)
capabilities 42–43, 51, 62
Cayman Trough 62–63,
photo insert
crew 43–44, 54, 92, 128
Cronkite aboard 100
East Pacific Rise 72–73,
photo insert
equipment 71, 73
Galápagos Rift 65–66, 67
Gulf of Maine 49
Mid-Atlantic Ridge 50
near calamities 57–59
Project FAMOUS 51,
54–59
refurbishment 48, 51
science community and
46, 84, 89
Scorpion expedition 135
St. Croix 92–93
Thresher expedition 135
Titanic expedition 128–
134, **photo insert**

Amagiri (ship) 232, 233, 235
Amphoras 146, 153, 193–194,
196–197, 214–215, 217–221,
271–272
Amundsen, Roald 227
Andrea Doria, S.S. 45, 194
Andrews, Thomas 101
Angus (camera sled)
capabilities 64, 94
Cayman Trough 64
East Pacific Rise 72–74
Galápagos Rift 65–66
Scorpion wreckage 104–105
Titanic expeditions 116–117,
133
Antarctica 227–228
Archimède (bathyscaphe)
51, 52–54, 58, 59
Argo (ROV)
Bismarck search 147–148,
154, **photo insert**
design and construction
93–94
funding 89, 153, 163
JASON Project 145
Mediterranean ship-
wrecks 145, 146
planning 83–84
Scorpion wreckage 95,
104–105
testing 163
Thresher wreckage 95,
96–98
Titanic search 96, 104,
106–110, 112–116, 118
Argus (imaging sled)
Black Sea expedition 224
Earhart search 300, 301

Eratosthenes Seamount
259
Gallipoli expedition 256
PT-109 search 235, **photo
insert**
Titanic exploration
244–246
Turkish Air Force jet
search 267, 269
Arizona, U.S.S. 206
Armor, Tim 182
Armstrong, Dennis 275–276
Ashe, Arthur 33–34
Astor, John Jacob, IV 250
Atlanta, U.S.S. 176
Atlantis II, R/V 126–134
Attention deficit/hyperac-
tivity disorder 288–289
ATV (*Advanced Tethered
Vehicle*) 206, 207–208

B

Ballard, Barbara *see* Earle,
Barbara
Ballard, Benjamin (son)
birth and childhood 188,
197, 199, 206, 229
Brooks talk 307, 308
dyslexia 289
Nautilus expeditions
255–256, 271–272,
photo insert
Ballard, Blake (grandson)
308
Ballard, Carol (sister-in-law)
204
Ballard, Carter (grandson)
308

Ballard, Chet (father)
  Ballard's childhood 13–16,
    19, 20–21
  at Ballard's wedding 170
  career 15–16, 109
  character traits 16, 17–18
  childhood 16
  death 204
  dyslexia 288
  East Pacific Rise expedi-
    tion 74
  later life 203–204
  relationship with Ballard
    203–204
  World War II 203
Ballard, Douglas (son)
  Black Sea expedition
    photo insert
  brother Todd and 143, 158
  character traits 289
  childhood 47, 76, 134–135,
    157–158, 170, **photo
    insert**
  children 308
  Colorado River rafting
    trip 142
  Queen Elizabeth 2 126
  Titanic discovery 119
Ballard, Emily (daughter)
  birth 200–201
  DEEP SEA Space 309
  dyslexia 289
  Earhart search 297, 303,
    304
  as *Nautilus* crew member
    photo insert
  Tolkien quotation 310
Ballard, Harriet (mother)
  Ballard's childhood 14–19
  on Ballard's Hatchville
    farmhouse 47
  on Ballard's use of sailor
    language 23–24
  at Ballard's wedding 170
  character traits 16–17, 18
  death 289
  Earhart as hero 293
  family background 17
  as housewife 48

later life 204
spelling ability 288
on *Titanic* discovery
  11–12, 126
Ballard, Jan (sister-in-law)
  170
Ballard, Jeff (nephew) 204
Ballard, Marjorie *see* Hargas, Marjorie
Ballard, Nancy Ann (sister)
  15, 18–19, 170, 204, 289,
  photo insert
Ballard, Richard (brother)
  on Ballard's luck 102–103
  at Ballard's wedding 170
  brilliance 15, 18, 28, 170
  character traits 17–18, 19
  childhood 13–18, **photo
    insert**
  death 289
  education 20, 28, 35
  father's later life 203, 204
  marriages 35, 38, 48, 76,
    160–161
  medical issues 20
Ballard, Robert
  ADHD 14, 19, 288–289
  ambition 27–28, 35
  athleticism 20
  calm under pressure 33
  childhood 13–21, **photo
    insert**
  childhood ambition to be
    oceanographer 15, 19–25
  determination and tenac-
    ity 16, 36, 305
  dyslexia 18, 19, 29, 47,
    285–290, 308–309
  education 27–30, 33–36,
    38, 46–49, 54, 59
  energy 43–44, 47, 55, 285
  entrepreneurial instincts
    17, 19, 83
  hyperfocus 55, 285, 303
  insecurity 285
  leadership 31–33, 34, 303
  luck 102–103, 140
  marksmanship 30, 33
  move to east coast 39, 41

as prankster 19
relationship with Barbara
  Earle 167–171, 197–199,
  photo insert
relationship with Marjo-
  rie Hargas 37–39, 48, 76,
  160–162
ROTC 30–34
time management and
  discipline 288–289
as visual thinker 30–31,
  285, 286
*see also specific people,
  places, topics, and
  vessels*
Ballard, Thomas (ancestor)
  137
Ballard, Todd (son)
  Bismarck search 147–148,
    153–155, **photo insert**
  character traits 142–143,
    146, 158–160, 289
  childhood 47, 76, 134–135,
    photo insert
  Colorado River rafting
    trip 142
  death 157–161
  Mediterranean scouting
    expedition 146
  salmon fishing in Canada
    143
  Titanic discovery 119
Bascom, Willard 81, 211–213,
  215, 221, 224, 226
Bass, George 222–223, 252
Bechtel, Riley P. and Susie
  258
Bell, Alexander Graham 45
Bell, Katy Croff 264–265,
  268–269
Ben (autonomous surface
  vessel) 300, 301
Berger, Lee 282
Big Events 80
Binkley, Johnny 17
Bioluminescence 56
Bismarck (German
  battleship)
  crew deaths 141

debris field 153–154
search for 144, 147–149,
153–155, **photo insert**
Black, Charlie 123–124
Black, Shirley Temple
123–124
Black Sea 211–226, **photo insert**
anoxic layer 81, 212–213,
221, 224, 225
Bascom's research 81,
211–213, 215, 221, 224
flood theory 213–215,
221–225
research funding 141, 239
scouting expedition
220–222
shipwrecks 12, 81, 140,
225–226, **photo insert**
Black smokers 72–75, 76,
photo insert
Boston Museum of Science
150
Boston Sea Rovers 44–45,
61, 222, **photo insert**
Bowen, Andy 93, 150, 176,
180–181
Bowen, Martin 104, 107, 117,
131, 152, **photo insert**
Bradt, Laurie 249, 252,
275–276
Bright Shark (Ballard) 214
Britannic wreckage 195
Broad, Kenny 282–284
Broderson, George 43–44
Brokaw, Tom 118
Brooks, Arthur 307–308
Brown, Betty Klenck 295,
298, 299
Bush, George W. 230, 238, 240

C

Californian (steamship) 122
Calypso (ship) 254
Cameron, James 200–202,
208
Campbell, Joseph 162–163,
167, 188, 307

Campbell, Kurt M. 294, 295
Canaanite amphoras
271–272
Canberra, H.M.S. 176, 179
Capable, U.S.N.S. 241
Carolyn Chouest, M/V 195
Carothers, Robert 238
Cattle drives 263
Cayman Trough 62–63, 64,
69–71, 76, **photo insert**
Charles, Prince of Wales
123–125
Chemosynthesis 68
Clams 66–68, 260
Clark, Eugenie 45
Claudius, Gordon 279
Claudius, Herbert 276–279
Clinton, Bill 238, 240
Clinton, Hillary 294–295
CNEXO (French National
Center for Exploitation of
the Oceans) 49–50
Connell, Hugh 189–192
Corliss, Jack 65, 66
Cortez, Sea of, Mexico 183
Cousteau, Jacques 44–45,
55, 223, 237, 254
Cousteau, Philippe 45
COVID-19 crisis 309
Crabs 268
Craig, Harmon 77
Cronkite, Walter 99–100
Crystal, Billy 136
Culberson, John 242
Cyana (submersible) 54, 59,
72–73
Cyprus 258–261

D

Dakar (submarine) 214
Dalton, Chad 158
Darwin, Charles 65, 165
DEEP SEA Space 309
Deep Tow (sonar vehicle) 65,
72, 73, 205
Deep Water, Ancient Ships
(Bascom) 212
Delta (mini-sub) 185–186

Dettweiler, Tom 93
Diamond, Neil 125
Diana, Princess of Wales
123–125
DiCaprio, Leonardo 200,
202
Disney, Walt 289
"Doctors on Call" program
261
Dodd, Chris 240
Dolphins 36–37, 38, **photo insert**
Dorman, Craig 164, 191
Douglas, Kirk 19–20
Drew, Christopher 90
Duck hunting 138, 139–140
Dyslexia
Ballard's 18, 19, 29, 47,
285–290, 308–309
The Dyslexic Advantage
(Eide and Eide)
285–290
Lubetkin and 303–304
multisensory museum
exhibits 309

E

Earhart, Amelia 293–305,
photo insert
Earle, Barbara
background 168, 191
character traits 180, 182,
183–184, 198
Connecticut home
199–200
Earhart search 297, 299,
302
friendships 308
Galápagos Islands 180,
182
Jackson Hole, Wyoming
263
Jason Project 164–167
Lusitania expedition 183–
188, **photo insert**
motherhood 184, 188, 197,
199, 200–201, 206, 289
National Geographic

Television specials 227
Odyssey Enterprises 170,
171
relationship with Ballard
167–171, 197–199, **photo
insert**
seaQuest DSV 184
South Pacific trip 173–180
Yorktown search 206
Earle, Bobbie 168–169
Earle, Harry 168–169, 202
Earle, John 173–174
East Pacific Rise 71–75,
photo insert
Edgerton, Harold 45
Edmond, John 67, 77
EDS (company) 144–145
EEZ (exclusive economic
zone) 273–275, 296, 305
Eide, Brock L. and Fernette
F. 285–288
Emery, Kenneth O. 43, 46,
49–50
Endurance shipwreck
227–228
Enterprise, U.S.S. 203
Eratosthenes Seamount
258–262, 271–273
Erciyes, Cagatay 268
Ertan, Gokhan 268, 269
Evans, Reginald 233, 235
Ewing, Maurice 51
Exclusive economic zone
(EEZ) 273–275, 296, 305

F
Fahey, John 230–231, 261
FAMOUS *see* Project
FAMOUS
Fort Lewis, Washington
31–33
Foster, Dudley 72, 129, 131
France
CNEXO 49–50
Titanic search 96, 99, 102,
103, 112–113, 120, 121,
127–128, 243
Trieste (bathyscaphe) 44

see also Project FAMOUS
Francheteau, Jean 51–52,
71–73
Francheteau, Marta 52
Franklin, Sir John 228
French-American Mid-
Ocean Undersea Study *see*
Project FAMOUS
Fundis, Allison 297, 300,
302–303, 304, 308, 309
Fye, Paul 43, 80–82

G
Gagosian, Robert 191, 192
Galápagos Islands 165–166,
180–183
Galápagos Rift 65–67, 72,
124, **photo insert**
Gallipoli, Battle of (1915)
255–257
Gasa, Biuku 234–235
Gegg, Steve 73–74
Giddings, Al 45
Gillespie, Ric 294–297, 299
Gimbel, Peter 45
Glenn, John 223
Glomar Explorer (ship) 81
Gokceada, T.C.G. 265, 270
Goodall, Jane 223
Gore, Al 201
Gray, Harry 200
Grayscout (ship) 230,
233–237
Greenert, Jonathan 270,
278–279
Gregg, Judd 238
Griffith, Dwaine 136, 170,
174
Grimm, Jack 83, 86–87,
94–95, 106, 133
Grosvenor, Gil 164, 169
Grosvenor, Melville Bell 45,
61
Guadalcanal, Battle of
(1942–1943) 173–180
Gustafson, Kirk 153, 155
Gustafson, Mike 163–164
Guven (boat) 221–224

H
Hamilton (schooner) 164
Hargas, Marjorie
Loch Ness monster trip 63
marital problems and
divorce 48, 76, 160–162
marriage to Ballard 37–39,
41, 47, 76, 134–135,
photo insert
Queen Elizabeth 2 126
Titanic discovery 119
Todd's death 157–158, 161
White House dinner 124,
125
Whitefish Lake, Montana
photo insert
Harris, Stu 93, 104, 110
Hart, Eva 243
Hayward, Thomas 85, 87,
88–89
Heirtzler, Jim 50
Hennes, Tom 200
Hercules (camera system)
design and construction
228, 239
Earhart search 300, 301,
302, **photo insert**
Eratosthenes Seamount
271–273
funding 229
Gallipoli expedition 256
Robert E. Lee expedition
277–278
Titanic exploration 242,
244–246, **photo insert**
Turkish Air Force jet
search 267–268, 269
Hershey, Bill 151–152
Hie, Fredrik 221
Hillary, Sir Edmund 223
Hitler, Adolf 276
Hohmann, Joe 44, 45
Hollings, Fritz 229, 237–238,
240
Hollis, Ralph 129, 133
Hollister, Charlie 137–138
Homer (remote-controlled
vehicle) 185

Hornet, U.S.S. 203
Howland, John 89
Howland Island 295–298,
305
Hubbs, Carl 24, 25
Huet de Froberville, Gérard
58
Hugo (ROV) 148–149,
151–152
Huizenga, Wayne and Marti
258
Hydrothermal vents 66–68,
71–76, **photo insert**

I

In Search Of (TV show) 64
Indianapolis, U.S.S. 228–229
Inouye, Daniel 240
The International Group for
Historic Aircraft Recovery
(TIGHAR) 294–296, 297
International Seapower
Symposium 84–85
Irresistible shipwreck 255,
256
Isis (shipwreck) 146, 149, 151,
153, 194, photo insert

J

Jackson Hole, Wyoming 263
James, Chris 186
Jannasch, Holger 71
Jason (ROV)
accidents 151–152
design and construction
83–84, 93–94, 148–149
funding 89, 148, 150
Isis expedition 151–153
Jason II project 164,
167–168
Jason III project 165–166
JASON Project 151–153
Lusitania wreckage 185,
186–187, **photo insert**
Mediterranean Sea 195,
196, **photo insert**
Phoenician shipwrecks
217–219

prototype (*see Jason Jr.*)
robotic hand 219
Sea of Cortez, Mexico 183
JASON Foundation for Edu-
cation 163–164, 182
Jason II 164, 167–168
Jason III 165–166, 180–183
Jason Jr. (*JJ*) 127–133, 135,
136, 180–181, **photo insert**
JASON Project
Bismarck search 147
development of 142, 144–
145, 150
end of 248
Eratosthenes Seamount
261–262, 271–273
films about 155
Florida Keys 199
funding 258
Hawaii 199
Isis expedition 150–153
Jason II 164, 167–168
Jason III 165–166, 180–183
science courses for
schoolchildren 141–142,
149–150, 152–153, 167–
168, 183
JJ see Jason Jr. (*JJ*)
Johnson, Osa and Martin
175
Joubert, Dereck and Bev-
erly 282

K

Kadioğlu, Yaşar 267, 269
Kaplan Fund 200
Karalekas, Spike 99–100
Karalekas, Tina 99
Kelly, Tim 149, 161, 164, 169
Kennedy, Caroline 230–231
Kennedy, Danny 230, 234
Kennedy, Jacqueline 229
Kennedy, John F. 29,
229–237
Kennedy, Max 231, 234, 236
Kennedy, Ted 229, 230–231
Keresey, Dick 235–236
King, Larry 201

Knorr, R/V 104–119, **photo
insert**
Kohler, Richie 277–279
Koll, Don 139–141, 143–144,
153, 258
Kollmorgen, Leland 88, 89
Kristof, Emory 62–63, 83,
103, 104, 108–109, **photo
insert**
Kumana, Eroni 234–235, 237

L

Laney Chouest (ship) 206
Lange, Bill 110
Lavoy, Peter 294
Le Pichon, Xavier 49, 52
Le Suroit (ship) 99
Lehman, John
Ballard friendship 99, 170
character traits 92
maritime strategy 84–86,
92–93, 136
Titanic search 92–93, 95,
121, 127
Levy, Leon 216
Little Hercules (camera) 224,
235, 236, **photo insert**
Littlehales, Bates 45
Lloyd's of London 145
Loch Ness monster 63–64
Logan, Lara 257, 277
Lord, Walter 101
Lubetkin, Megan 303–304,
308, 309
Lusitania wreckage 173,
183–188, **photo insert**
Luyendyk, Bruce 50–51
Lyman Allyn Art Museum,
Connecticut 309

M

Mabus, Ray 279
MacInnis, Joe 45
Maine, Gulf of 46–49
Map 8–9
Marden, Luis 45
Maritime Management 253
Marquest (company) 136, 163

Marschall, Ken 186, **photo insert**
Mashantucket Pequot Tribal Nation 200
Matthias, Paul 187
Mayer, Larry 309
McCann, Anna 153, 194–197
Mead, Margaret 30
Meagher, Janice 252, 275–276
Medea (ROV) 152
Mediterranean Sea research
  amphoras 146, 153, 193–194, 196–197, 214–215, 217–219, 271–272
  ancient trade routes 12, 136, 193, 197, 259, 260–261, 272
  Bass's research 222
  downed bombers 43
  Eratosthenes Seamount 258–262, 271–273
  funding 141
  JASON Project 145
  scouting expedition 145–147
  shipwrecks 140, 141, 146, 193–197, 214–220, 239, **photo insert**
Michel, Jean-Louis
  Project FAMOUS 51
  Titanic search 95–96, 102–104, 108–110, 112–113, 115, 120, 121, 127, **photo insert**
Mid-Atlantic Ridge 49–59
Mid-ocean ridge 85
Mindell, David 195–196, 207
Mooney, Brad **photo insert**
Morse, Robert W. 77, 84
Moyers, Bill 162–163
Mussels 66
Mystic Marinelife Aquarium, Connecticut 189–192, 198, 200, 202, 249, 252

N

National Geographic Channel 244, 246

National Geographic magazine 45, 61, 63–64, 76, 127, 134
National Geographic Society
  Centennial Award 223
  Earhart search 296–297, 299–300, 305
  Explorers Festival 281–285, 290, 292
  Genographic Project 283
  JASON Foundation 164
  photographers 45
  Pristine Seas 282
  Titanic discovery 120, 121–122, **photo insert**
  Titanic videos 135–136
National Geographic Television
  On Assignment 161–162
  Bismarck search 141, 147–148, 149, 153
  closure of studio 261
  Endurance expedition 227
  Gallipoli shipwrecks 255, 256–257
  hydrothermal vents 76
  JASON Project 155
  The Lost Fleet of Guadalcanal 173, 178, 179
  Lusitania 173
  National Geographic Explorer 161
  Nazi Attack on America 276
  Phoenician shipwrecks 218
  The Search for Kennedy's PT-109 229–231, 237
  Titanic expeditions 134, 141
  Yorktown search 206
National Oceanic and Atmospheric Administration (NOAA) 237–238, 240, 241, 257–258, 273–274, 305
National Science Foundation 62, 84, 89

Nautile (French mini-sub) 243
Nautilus, E/V
  acquisition (as *Humboldt*) 249–251
  crew 253, 266, 268–269, 271, **photo insert**
  Earhart search 296, 300–305, **photo insert**
  EEZ mapping 275, 296
  equipment 253, 254, 257, 300
  Eratosthenes Seamount 258–262, 271–273
  funding 257–258, 274, 275
  Gallipoli shipwrecks 255–257
  lead scientist 264–265, 268–269
  need for 247–249, 254
  Robert E. Lee expedition 277–279
  telecommunications 261, 262
  Turkish Air Force jet search 263–271
  Turkish base 252–253, 254–255
  Walking with the Ancients project 284
Navy *see* U.S. Navy
Newman, Jim 249–250
Nichols, Guy 164
Nichols, Roger L. 150
Nielson, Terry 84
A Night to Remember (Lord) 101
Nikumaroro island 294–301, 305, **photo insert**
Nimitz, Chester 202–203
Nimoy, Leonard 64
Nixon, Richard 189–190, 237
NOAA 237–238, 240, 241, 257–258, 273–274, 305
Noah's Flood (Ryan and Pitman) 213–214, 215, 221–225
Noonan, Fred 295, 298–299, 300

Norris, Ken 36, 38
Norris, Robert 22, 28, 35, 36
North American Aviation 29
Northern Horizon (ship)
 184–188, 217–220, 224–226
Norwich City, S.S. 295–296,
 298, 300, 301, 302, **photo
 insert**
NR-1 (submarine) 194–196,
 199, 214, **photo insert**

O

Ocean
 atmospheric pressure
  56–57
 government funding for
  exploration 229, 237–
  238, 240–242, 273
 layers 55–56
 seafloor mapping 49, 50
 seawater chemistry 74–75
 terrain warfare 85, 87–89,
  91
Ocean Exploration Trust
 (OET) 252, 276, 297, 309
Oceanic Institute, Hawaii
 36–37, 38
Odyssey Enterprises 170,
 171, 173–180
Office of Naval Research 39,
 41–42, 84, 88, 89
Offinger, Cathy 64, 150–151,
 181, 208, 230
Okeanos Explorer (ship) 241,
 258, 274
Ontario, Lake 164, 167
Operation Drumbeat 276
Orca, R/V 22–23
O'Reilly, David and Joan
 258

P

Patricelli, Robert 309
PC-566, U.S.S. 276–279
Pearl Harbor, attack on
 202–203, 206
Pellegrini, Ann 136, 145
Pelli, César 200

Phillips, Brennan 268, 269
Phoenician shipwrecks 214–
 220, 239
Piccard, Jacques 44
Pitman, Walter 213–214, 215,
 221–225
Plate tectonics
 Cayman Trough 62–63
 Galápagos Rift 65
 Gulf of Maine 46–49, 59
 Mediterranean Sea 259,
  260
 Mid-Atlantic Ridge 49,
  52–53, 57–59
 National Geographic
  Television special 76
Press, Frank 50–51
Project FAMOUS 49–59, 75
PT-109 229–237, **photo
 insert**

Q

Queen Elizabeth 2 (ship) 126

R

R Lazy S Ranch 263
Rainnie, Bill 43, 46
Rakestraw, Norris 21
Raney, Chris 175, 176
Ravenscroft, Alan 83
Reagan, Nancy 99, 125
Reagan, Ronald 84–85,
 92–93, 123–125, 127, 139
Rechberger, Kristin 282
Rechnitzer, Andrew 39
Restless M (Australian sup-
 ply boat) 174–179
Revelle, Roger 21, 27
Ricciardone, Francis J., Jr.
 263–264
Richard, Chas 214
Rickover, Hyman G. 88,
 91–92
Ridder, Dale 235–236
Robert E. Lee (ship) 276–279
Roman, Chris 268
Ronald H. Brown (NOAA
 flagship) 244–246

Rose, Lana 29, 34
Ryan, William 86, 107, 213–
 214, 215, 221–225

S

Sagan, Carl 123
Sala, Enric 282
Sauder, Eric 186
Scalli, Frank 44–45
Scheider, Roy 184
Schempf, Hagen 159–160,
 188
Scorpio (deep-towed side-
 scan sonar system) 179
Scorpion, U.S.S. 69, 90–91,
 95, 102, 104–106, 128, 135,
 136
Scourge (schooner) 164
Scripps Institution of
 Oceanography
 Ballard and 20–25, 34–
  36
 Capricorn Expedition 211
 Deep Tow 205
 East Pacific Rise 71–75
 Galápagos Rift 65
 research facilities 21
 WHOI rivalry 39, 42, 43,
  73–74, 81, 93
Sea Cliff (submersible) 179
Sea Life Park, Hawaii 36–37,
 photo insert
Sea lions 180
Sea Research Foundation
 192
Sea Rovers 44–45, 61, 222,
 photo insert
Seamounts 258–262,
 271–273
Seaprobe (ship) 81–82,
 212–213
seaQuest DSV (TV show) 184
Sender, Karen 207
Shackleton, Sir Ernest
 227–228
Shelley, Mark 187
Shepard, Alan 125
Shepard, Francis 22, 43

Shipworms 212

Sigurdsson, Haraldur
165–166

Silva, Gene 88

Simon, William 200

Sims, Neil 291

Sinop, Turkey 221–222, 224, 239

Sinop D shipwreck 225–226, 239

60 Minutes (TV show) 257, 277

Skerki Bank expedition 146, 195–198

Slickensides 260

Smart, Claire 279

Smith, Edward 100–101

Solomon Islands 173–180, 229–237

Sontag, Sherry 90

South Pole 227–228

Spielberg, Steven 184

Spiess, Fred Noel
Ballard and 34–35, 76, 77, 191
Deep Tow 205
East Pacific Rise 72, 74
Titanic search 86, 107
WHOI rivalry 93

Spradlin, Diane 144

Spurr, Cyril 186–187

Squires, Bob 93

St. Croix 92–93

Stager, Larry 214, 216–220, 272

Stanford University 65–66, 75–76, 83

Star Hercules (ship) 150–151

Starella, M/V 145–148

Steele, John
conflicts with Ballard 120–123, 191
Jason expedition 150
Titanic and 83, 111, 118, 120–121, 127–128
WHOI Center for Marine Exploration 126, 127, 137, 144

Steinbeck, John 21, 211–212

Stevens, Ted 240–241

Submersibles, manned
limitations 83–84, 128, 131–132
North American Aviation 29
Project FAMOUS 52
scientists' disdain toward 50–51

Submersibles, robotic 71, 89

Sullivan, Walter 58, 62–63

Sunfish (autonomous underwater system) 284

Surgi, Bill 209

Syria 263–267

T

Tantum, William H. 80, 83, 101–102, 133

Taylor, Rod and Valerie 45

TBS *see* Turner Broadcasting

Tectonic plates *see* Plate tectonics

Teredinidae (shipworms) 212

Thayer, Jack 100, 101–102, 106, 111, 116, 287

The International Group for Historic Aircraft Recovery (TIGHAR) 294–296, 297

Thresher, U.S.S. 89–90, 95–98, 105–106, 135, 136

Thunman, Ronald 87–92, 95

Titanic (film) 45, 200–202, 208

Titanic, R.M.S.
artifacts 132, 242–244, 245–246, **photo insert**
Atlantis II visit (1986) 126–134, **photo insert**
Ballard's books about 134, 144

Ballard's pipe dream 45, 79–83, 91–93, 95–96, 100–110

Big Events search 80

debris field 106–107, 110, 112, 132, 287

French search 96, 99, 102, 103, 112–113, 120, 121, 127–128, 243

Grand Staircase 91, 116, 127, 130–131

grandeur 79, **photo insert**

Grimm/Spiess/Ryan search 86–87, 106–107

Grimm's search 83, 86–87, 94–95, 133

hull 112–117, 133–134

location 95, 103

media frenzy 11, 118–122, **photo insert**

model 200

press conferences 120, 121–122, 134

reactions to discovery 11–12, 109–111, 115, 118, **photo insert**

return visit (2004) 242, 244–246, **photo insert**

status boost from discovery 123–126, 134, 136–137

survivors 243

tragedy 80, 100–102, 111, 122, 133–134, 250

turf wars 120–121, 127–128

WHOI search 80–83

Titanic Historical Society 80

Titanic's Lost Sister (PBS show) 195

Tolkien, J.R.R. 310

Torgerson, Paul 163–164

Travolta, John 125

Trieste (bathyscaphe) 44

Trieste II (bathyscaphe) 69–71, 96, 105

Triumph shipwreck 256
Tube worms 66–67, 68, 71,
124, 260, **photo insert**
Turanli, Tufan 252, 265
Turkish Air Force jet
search 263–271
Turner, Ted 149
Turner Broadcasting (TBS)
134, 141, 149, 151, 167–168,
227
*Twenty Thousand Leagues
Under the Sea* (film)
19–20
Tzimoulis, Paul 45

U
U-166 (Nazi submarine)
276–279
Uluburun shipwreck 222
University of California at
Santa Barbara 27–30, 34
University of Hawaii at
Manoa 36, 38, 205, 206
University of Rhode Island
46–48, 54, 238–239, 261,
274–275
U.S. Divers 44–45
U.S. Navy
Ballard's classified work
136
Deep Submergence Sys-
tems group 89
Guadalcanal anniversary
174, 179
International Seapower
Symposium 84–85
ROV funding 148, 153,
163
Submarine Development
Group One 128, 179,
205–206
Titanic expeditions
91–93, 95, 104–118, 121,
127–134
undersea terrain warfare
85, 87–89, 91
Yorktown search 205–
206, 207–208

see also *NR-1*; Office of
Naval Research;
PT-109; *Scorpion*;
Thresher

V
Van Andel, Tjeerd Hendrik
66, 75, 77
Verd, George 89
Verne, Jules 19–20, 310
Vestron Video 135–136
Vietnam War 30, 38, 39
Viola, Vincent 249–252,
264–266, 270, 287
Vitulli, Marco 139–140,
143–144, 153

W
Walking with the Ancients
project 283–284
War of 1812 164
Waterman, Stan 45
Weber, Chris 148, 180, 297
Weicker, Lowell 189–192
Wellington, Jerry 166, 183
Wells, Spencer 283
White, Shelby 216
White House dinner
123–125
Whitefish Lake, Montana
76, 134–135, 157–158, 160,
162, **photo insert**
Williams, Robin 136
Winslet, Kate 200, 202
Wolf, Frank 241–242
Woods Hole Oceano-
graphic Institution
(WHOI)
Atlantis II 126–134
Ballard as Office of Naval
Research liaison to 39,
42–44
Ballard's career 46, 48,
62, 64, 71, 75–77, 190–
192, 199
Center for Marine Explo-
ration 126, 127, 137,
144

Deep Submergence
Group 43
Deep Submergence Lab-
oratory 93–94, 104
funding 137–139, 141, 144,
190
Galápagos Rift 65
Jason sponsorship 150
Scripps rivalry 39, 42, 43,
73–74, 81, 93
Titanic expeditions
80–83, 119–121, 132,
134, 190
see also Project FAMOUS

Y
Yeager, Chuck 15, 109
Yoerger, Dana 93, 108, 150,
151–152
Yorktown, U.S.S. 197–198,
202–209, photo insert
Young, Carole 144
Young, Earl 64, 104, 113–
114, 117
Young, Jim 144
Yunck, Billy 153, 155

ILLUSTRATIONS CREDITS

Photos courtesy Robert Ballard unless otherwise noted.

ABOUT THE AUTHORS

Best known for discovering hydrothermal vents, the sunken R.M.S. *Titanic,* the German battleship *Bismarck,* and many other ancient and modern shipwrecks around the world, Robert D. Ballard has conducted more than 150 deep-sea expeditions and is a pioneer in the development of advanced deep submergence and telepresence technology. Since 2008, he has managed E/V *Nautilus,* his flagship for exploration, operated by the Ocean Exploration Trust and funded in part by NOAA. Among his many honors, he holds the Explorers Club Medal, the National Geographic Hubbard Medal, and the National Endowment for the Humanities Medal. He lives in Lyme, Connecticut, with his wife, Barbara.

—

An investigative reporter and editor at the *New York Times* for 22 years, Christopher Drew co-wrote the best-selling book *Blind Man's Bluff: The Untold Story of American Submarine Espionage.* He also has worked for the *Chicago Tribune* and the *Wall Street Journal.* He now holds the Fred Jones Greer Jr. endowed chair at the Manship School of Mass Communication at Louisiana State University in Baton Rouge, Louisiana. He lives in Baton Rouge with his wife, Annette.

NATIONAL GEOGRAPHIC

THE NATIONAL GEOGRAPHIC SOCIETY INVESTS IN BOLD PEOPLE WITH TRANSFORMATIVE IDEAS WHO ILLUMINATE AND PROTECT THE WONDER OF OUR WORLD.

Since 1888, the National Geographic Society has driven impact by identifying and investing in a global community of Explorers: leading changemakers in science, education, storytelling, conservation and technology. National Geographic Explorers help bring our mission to life by defining some of the most critical challenges of our time, uncovering new knowledge, advancing new solutions, and inspiring transformative change in our world.

One of the most revered members of this community is Explorer at Large Robert Ballard, a renowned expert in ocean exploration and underwater research. With a career spanning five decades, Ballard is a testament to an unyielding spirit of exploration.

To learn more about the Explorers we invest in—like Ballard—and the efforts we support, visit natgeo.org.

ABOUT THE NATIONAL GEOGRAPHIC SOCIETY

The National Geographic Society is a global nonprofit organization that uses the power of science, exploration, education, and storytelling to illuminate and protect the wonder of our world. Since 1888, National Geographic has pushed the boundaries of exploration, investing in bold people and transformative ideas, providing more than 14,000 grants for work across all seven continents, reaching three million students each year through education offerings, and engaging audiences around the globe through signature experiences, stories, and content. To learn more, follow us on Instagram, Twitter, and Facebook at @InsideNatGeo.